Gaslighted

Gaslighted

How the Oil and Gas Industry
Shortchanges Women Scientists

Christine L. Williams

UNIVERSITY OF CALIFORNIA PRESS

University of California Press
Oakland, California

Library of Congress Cataloging-in-Publication Data

Names: Williams, Christine L., 1959– author.
Title: Gaslighted : how the oil and gas industry
  shortchanges women scientists / Christine L.
  Williams.
Description: Oakland, California : University of
  California Press, [2021] | Includes bibliographical
  references and index.
Identifiers: LCCN 2021016032 (print) | LCCN 2021016033
  (ebook) | ISBN 9780520385276 (hardback) |
  ISBN 9780520385283 (paperback) |
  ISBN 9780520385290 (ebook)
Subjects: LCSH: Oil industries—United States—
  Employees. | Gasoline industry—United States—
  Employees. | Women earth scientists—United States.
  | Sex discrimination against women—United States.
Classification: LCC HD8039.O52 U69 2021 (print) |
  LCC HD8039.O52 (ebook) | DDC 331.4/8655—dc23
LC record available at https://lccn.loc.gov/2021016032
LC ebook record available at https://lccn.loc.
  gov/2021016033

Manufactured in the United States of America

30 29 28 27 26 25 24 23 22 21
10 9 8 7 6 5 4 3 2 1

CONTENTS

ILLUSTRATIONS

FIGURES

TABLES

ACKNOWLEDGMENTS

I am grateful to the women of PROWESS who inspired this study. Although they sought me out for my expertise, I ended up learning a great deal from them. I also thank the geoscientists and engineers who agreed to talk to me over the years. I sincerely appreciate their willingness to share their insights and experiences with me during difficult times.

My debt to Chandra Muller is huge. She convinced me to take on this project and was enthusiastic from the start. I have enjoyed our collaboration over the decade, as well as our friendship. A number of graduate students worked with us on the design and development of this project, including Jessica Dunning Lozano, Kristine Kilanski, Samantha Simon, and Amanda Bosky. I thank them for their many contributions to the project.

I presented parts of this work at the American Sociological Association meetings, at the Southern Sociological Society, and at a conference on the future of work in Queensland, Australia. Thank you to Toni Calasanti, Michelle Brady, and Gillian Whitehouse for arranging those visits and for their valuable feedback on my work. Parts of chapter 5 appeared as "The Gender of Layoffs in the Oil and Gas Industry," in *Research in the Sociology of Work* (2017), in a special volume edited by Arne Kalleberg and Steven Vallas.

I am grateful for the helpful and encouraging feedback I received from brilliant sociologists Allison Pugh, Adia Harvey Wingfield, Steven Vallas, Jane Collins, and Jennifer Glass. Martin Button helped me figure out what I wanted to say and how best to say it. Profound thanks to Naomi Schneider, executive editor at UC Press, for her support from the very beginning.

As most of this book was written during the pandemic, I am grateful for Zoom. I looked forward to weekly family gatherings with my parents, Bunnie and Clyde Williams, my sisters and their partners, Cathie and YT Martinez, and Karrie Williams and Mark Jakusovszky, and my nieces Colleen and Claire Hunt. They cheered me every step of the way. While finishing the book, Fem(me) Sem meetings with graduate students Cait Carroll, Thatcher Combs, Shannon Malone Gonzalez, Katie Rogers, Patrick Sheehan, and Erika Slaymaker reminded me why sociology matters. Weekly happy hours with Jim Jasper, Mary Waters, and Deb Umberson kept me, well, happy. And last but never least, thanks as always to Martin, for making everything easier.

CHAPTER ONE

Gender, Geology, and the Oil and Gas Industry

In November 2008, I received an email from the Professional Organization of Women in Earth Sciences, or PROWESS. They needed a statistical consultant and they found me, a University of Texas sociologist who specializes in the study of gender and work. Members of PROWESS had just completed a survey of women geoscientists to find out why they were leaving their lucrative jobs in the oil and gas industry. They needed an expert to help them analyze the results.

Unbeknownst to them, I had no background analyzing survey data.

Unbeknownst to me, I was about to embark on a 10-year journey to answer their question.

The first thing I did when I received the email was to forward it to my colleague Chandra Muller, who really is an expert in survey data analysis. She quickly ascertained that their survey could not address the questions they were asking. Geoscientists had designed the survey after all, not social scientists. They omitted key questions—like "gender," for example—and they did not keep track of who responded to the survey. Nevertheless, I was intrigued and got caught up in their quest for answers. The women of PROWESS had opened up a fresh portal into some of the most enduring puzzles in the sociology of gender.

Why have some industries made strides toward gender
   equality while others remain stubbornly male-dominated?
Why is it taking so long for women to make inroads into
   scientific careers?
Why do high-paying jobs employ so few women?
What blocks women's ascent to corporate power?

Oil and gas is one of the largest, most lucrative, and most politically powerful industries in the world, yet it employs very few women. The industry lags behind virtually all others in measures of gender diversity. Women make up one-fifth of the industry's overall labor force, and they hold fewer than 15 percent of technical roles and very few leadership positions. In 2018, only one woman occupied the CEO position in a publicly traded oil and gas company, Vicki Hollub at Occidental Petroleum (Catalyst 2019; IHS 2014; Rick, Martén, and Lonski 2017).

The mission of PROWESS was to change that. PROWESS was a "special interest group" of the American Association of Petroleum Geologists, or AAPG. AAPG is a professional association with over 35,000 geoscientists working for the oil and gas industry in over 130 countries. On the AAPG website, the PROWESS mission was stated as follows:

> The mission of PROWESS is to increase participation and advancement of women in Earth sciences and the Energy Industry, with an emphasis on education, outreach, support, leadership development, and ultimately retention. PROWESS will interact

with women in Earth Science, their male peers and employers, educational institutions, and professional societies to accomplish this mission.[1]

AAPG cultivates close ties to the "energy industry" (a euphemism for oil and gas). Exxon/Mobil, Chevron, Shell, and British Petroleum—the so-called "super majors"—sponsor the Association, along with a cascading array of national companies (Saudi Aramco, Pemex), independent producers (Conoco-Phillips, Marathon, Occidental), and service companies (Schlumberger, Bechtel). These corporations and many others exhibit at the AAPG annual conventions, where they market their services and recruit geoscientists to work for them.

These companies employ geoscientists to find oil and gas. Geoscientists use computer modeling, seismic data, and historical records to predict the size and location of oil and gas deposits in the earth. In the big companies, they typically work on teams with engineers, who design and build the actual rigs. Geoscientists decide where to drill or hydraulically fracture the earth, but they rarely travel to oil fields or offshore drilling platforms. Most geoscience professionals are office-based, and in the United States, most work out of corporate headquarters in Houston, Texas, the nation's unofficial oil capital.

To become a geoscientist in the major oil and gas companies requires a master's degree. Around 40 percent of those receiving these degrees today are women. This is down from 45 percent in 2006, but it is still one of the highest percentages among the science majors. Corporations have committed resources to cultivating girls' interest in science. A recent effort, called "She Can STEM," targets 11-to-15-year-old girls with upbeat images of women engineers and scientists. General Electric, Google,

IBM, Microsoft, and Verizon, and a host of nonprofits, are collaborating with the Ad Council on this multimedia ad campaign. The *New York Times* quotes Ad Council CEO Lisa Sherman about the initiative:

> When girls don't feel encouraged and empowered in STEM, we see serious consequences not only for girls and women, but also for the future of innovation in our country. If we want women at the forefront of the next generation of STEM leaders, we must show young girls that it is possible. (Levere 2018)

Cultivating girls' interest in science and increasing the number of women STEM graduates are also long-term goals of federal education policy. The National Science Foundation (NSF) identifies the "leaky pipeline" as a cause of women's underrepresentation in science and engineering disciplines. Many young women arrive at college interested in science, but they eventually turn away, or "leak," from these programs, for a variety of reasons. Studies sponsored by the NSF identify several factors—including a lack of women role models, biased college curricula that reflect stereotypical masculine interests, and sexist cultures in academic departments—for young women's declining interest in STEM.

Compared to other STEM fields, the geosciences have fewer obstacles to women. Like biology, geology is a success story when it comes to increased gender equality in college enrollment and graduation. However, this is not the case for racial/ethnic diversity. Few of those graduating with geoscience degrees are from racial/ethnic minority groups; in fact, the geosciences have the highest percentage of white students among all the science and engineering majors in the United States (Wilson 2016:47). Thus the vast majority of the women graduates in the geosciences are white women.

The experiences of these white women provide a window to understanding how the oil and gas industry is shaped not only by gender, but also by race. White women have access to the profession of geology and job opportunities in oil and gas companies because of their race privilege, but the stereotypes of white femininity constrain their career development, just as in other STEM professions (Alfrey and Twine 2017).

For these women, attrition occurs *after* beginning their jobs. Those hired into the major oil companies do not last long in the industry. This is unfortunately the case in all the sciences: Within five years, about half of women scientists (compared to a third of men scientists) leave their careers and switch to nonscience fields (Glass et al. 2013). The PROWESS survey intended to find out why women geoscientists were leaving the oil industry. Granted their significant education investment and the lucrative salaries they were leaving behind, it was a good question.

CHERCHEZ LA FEMME

The PROWESS survey received the endorsement of the AAPG's Corporate Advisory Board, which consists of oil industry executives. The survey was sent to the entire 35,000-member organization, with instructions that only women should answer it. Over 2,000 people responded, including 500 who volunteered for a follow-up interview. The enthusiastic response indicated that PROWESS was definitely on to something important. Everyone agreed that women were leaving the industry and no one knew why.

After 10 years of researching this question, I think I now have the answer. Women leave the oil and gas industry because they are forced out.

This answer finally dawned on me in the eighth year of my research, when the price of oil plummeted. As chance would have it, I was studying the industry during one of its periodic downswings. Oil prices hit a high mark of $110/barrel in 2014. Skyrocketing prices were fueled in part by the "peak oil" hypothesis—the popular belief at the time that world oil production had reached its maximum and would inevitably decline until the earth's oil supply was exhausted. That very year, the fracking boom started and mammoth oil reserves were found in the United States. By 2016, the price of oil had fallen to $24/barrel, a 78-percent decline.

In 2016, nearly all of the major companies were laying off workers, including the company I studied. Hiring and firing in this industry track the price of oil: when oil prices are high, the industry goes on a hiring spree, offers signing bonuses, and implements retention and career development programs. When the price of oil drops, layoffs occur, assets are sold, and companies close down or position themselves for acquisition.

Layoffs are becoming increasingly common in many industries, not only in oil and gas. In the United States, companies face virtually no restrictions on laying off employees. The only federal law restricting employers is the Worker Adjustment and Retraining Notification Act, otherwise known as the WARN Act, which requires large companies to give advanced notice to employees about impending layoffs. Without any other laws to restrain them, major corporations in the United States routinely lay off workers during economic downturns; they also do so when economic conditions are good in order to strengthen stock prices and boost profits (Jung 2017; Vallas 2011).

Layoffs differentially impact women of all races, as well as racial/ethnic minority men (Byron 2010; Kalev 2020). These groups seem to be targeted by employers when companies downsize. In her study of over 800 US companies, economic sociologist Alexandra Kalev (2014) found that downsizing can increase the percentage of white male managers as much as 10 percent, while decreasing the percentage of white women and men and women of color by 22 and 17 percent, respectively.

The oil and gas industry is notorious in this regard. Following one company over a cycle of boom-and-bust enabled me to observe the impact of the downturn on geoscientists in the industry. Both men and women were losing their jobs, but women were bearing the brunt of the layoffs. Consistent with the national trends (Glass et al. 2013), a third of the men and half of the women exited the company over the course of this study. After years of education policy and diversity campaigns encouraging women to pursue scientific careers, the industry was kicking them out.

In this book I explore what this employment instability looks like from the points of view of scientists in the oil industry. I collected their personal narratives over the period of boom-and-bust for insight into the organizational processes that produce this gendered trend. The survey data reveal that layoffs affected men and women differently, but the interviews help us to understand the organizational forces that generated this outcome, as well as to explain what downsizing felt like for the people going through it. I talked with women geoscientists over the course of three years to understand how their company squeezed them out.

Women geoscientists in the oil and gas industry make up a tiny number of elite professionals. They are privileged on every dimension—race, education, income—except for gender. But despite their rarity, I believe that their experiences can illuminate why equality in the corporate world remains an elusive goal. An industry that only admits women who are white, and then targets them for layoffs, will not make progress achieving diversity. In the cyclical oil and gas industry, disproportionately laying off the women means that companies can revert to virtually all-white male bastions after every downturn. As layoffs become an ever more accepted and normal business practice throughout the economy, this form of discrimination will spread unless new rules are implemented to prevent it.[2]

To reach this conclusion, I have used almost every tool in the sociologists' toolbox. I conducted almost one hundred in-depth interviews. Working with my colleague Chandra Muller, I helped design and administer a longitudinal survey of the multinational oil and gas company that we call "GOG," or Global Oil and Gas (not its real name). I attended conferences and networking events around the country. I visited geoscientists at work in some of the world's largest oil and gas companies. I have been an invited speaker at the industry-sponsored Women's Global Leadership Conference, and at the professional meetings of petroleum geologists and geophysicists, where I have shared my findings and received feedback along the way.

Looking back at my work over the decade, it fascinates me that the answer to the question "Where are the women?" eluded me. Even some of the PROWESS women had been laid off previously in their careers. Why couldn't I see what was plainly in front of me from the very beginning? From my vantage point today, I feel like I was gaslighted.

Gaslighting is usually understood to be a form of psychological manipulation and emotional abuse in intimate relationships. The term comes from the name of a classic film in which a man convinces his wife to question reality and her own sanity. In this book, I argue that organizations can also engage in gaslighting. Organizational gaslighting is when companies intentionally deny the facts and blame others for the problems they generate. Corporations attempt to puff up their own image while denying evidence of their malfeasance, enabling them to escape culpability for the systemic inequalities they produce. For instance, they commonly use these tactics to make it appear that they support diversity:

- State a commitment to diversity in their mission statements
- Feature images of men and women from different racial/ ethnic backgrounds in publicity and advertisements
- Donate money to organizations or programs promoting equality that do not interfere with or challenge their normal business operations
- Implement diversity programs, such as unconscious bias training, that do not alter the composition of the workforce

The white male–dominated oil and gas industry does all of these things. Companies claim that they value diversity, but the employment policies they implement to achieve greater diversity do not disrupt the systemic sexism and racism, and other forms of social inequality, that are built into their organizations. Instead, they attempt to throw critics off the scent. This is the essence of organizational gaslighting. One of the goals of this book is to understand why I was misled and to encourage others to recognize organizational gaslighting when it happens to them.

In the early stages of this project, I spent my time looking for nuanced and subtle processes that reproduced the white male domination of the industry. I focused on how annual evaluations, promotion criteria, and organizational charts favored these men and qualities associated with elite masculinity (Acker 1990). I also explored the ways in which women might be holding themselves back. This study overlapped with "Lean In," a popular corporate-sponsored leadership program that encourages women to overcome their internalized sexism (Sandberg 2013). Thus, I wondered: Were women less motivated than men? Did they hold themselves back from pursuing management positions? Did their ambitions at work collide with their desired level of engagement in family life?

These are all important factors for understanding male-dominated industries. But these subtle processes are not what drive women away. I am now convinced that women do not leave high-paying professional careers unless they are forced out.

SEXISM AND THE OIL CURSE

Whenever I broach the subject of sexism in the oil and gas industry, people roll their eyes at me. Well, yes, of course, the oil and gas industry is sexist. "It's a man's world," they remind me. Drilling for oil is dirty, physical, dangerous work. The global hunt for oil is a lawless search for pirates' treasure for which women are neither suited nor welcome.

Few other industries are so closely identified with sexism. Those that are have been called to account in recent years, thanks in part to the #MeToo movement. In the last decade, we have seen such white male bastions as Hollywood, technology, and academia challenged on their abysmal records on diversity. But not so the oil and gas industry, which is headed by a global elite that is virtually all male (see figure 1).

Sometimes the industry is held accountable for the harm it causes to women *outside* the industry, particularly to women in the global South. In 2013, the World Bank published a study that found that women in developing countries rarely benefit, and

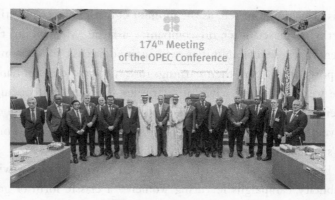

Figure 1. Delegates at the 2018 OPEC (Organization of Petroleum Exporting Countries) Conference, with representatives from the 15 Member Countries: Algeria, Angola, Congo, Ecuador, Equatorial Gabon, Guinea, Iran, Iraq, Kuwait, Libya, Nigeria, Qatar, Saudi Arabia, United Arab Emirates, and Venezuela.

more often suffer, when oil is discovered in their communities. Wherever they go, oil companies wreak environmental, economic, and social havoc, and women bear the worst of this damage. According to the World Bank, these include "pollution, land loss, rising prostitution, and alcohol consumption, and the de facto exclusion of women from consultations, decision making, and access to new income streams entering the household" (Scott et al. 2013:13). The report, based on research conducted in Azerbaijan, Peru, and Papua New Guinea, found that only men benefited from the few economic opportunities made available to them by oil companies, while women were left to fulfill their traditional tasks—feeding and caring for their families—under worsened conditions. The report found that oil companies depleted the supply of potable water, arable lands, and breathable air, making it more difficult for women to maintain their families' well-being. Women's reproductive health in these areas was also compromised, as the number of birth defects and still-births increased after oil companies arrived. Meanwhile, the men in these communities who gained access to economic opportunities spent their earnings in ways that did not enhance the welfare of their wives and families. The authors conclude their report by urging oil companies to invest more in local development and empowerment projects with a focus on women's needs (see also Perks and Schulz 2020).

For its part, the industry responded to this criticism with publicity campaigns featuring women—a classic form of gaslighting. After the massive oil spill caused by the Deepwater Horizon explosion in the Gulf of Mexico, British Petroleum put spokeswoman Iris Cross in their advertisements to emphasize their trustworthiness to clean up the catastrophe and restore the community. (See figure 2.) Chevron responded to the criticism with a publicity campaign highlighting their positive impact on women around the world, called "We Agree." A sample of these are shown in figure 3. These advertisements emphasize the positive impact of the oil industry, sidestepping the damage it wreaks on women throughout the world. They feature women representing small business owners, local community members, and aspiring leaders, attesting to the company's support for women both inside and outside the industry. These ads were immediately criticized by environmental groups, including the Rainforest Action Network, and called out as a form of organizational gaslighting called "greenwashing," a term that refers to deceptive marketing practices intended to give the false impression that a company's products and practices are environmentally safe.[3]

Toxic industries often use women spokespersons in their greenwashing campaigns because of their cultural association with nurturing (Bell, Fitzgerald, and York 2019)—a supposedly

Figure 2. British Petroleum Advertisement after the Deepwater Horizon disaster in the Gulf of Mexico.

innate difference from men that serves to reassure consumers that the corporation cares about the environment. The fact that these particular ads feature women of color is not by chance. Women of color do even more work for the industry by representing a commitment to social justice, a "racial task" that manages to simultaneously mask, normalize, and justify the whiteness of the industry (Wingfield and Alston 2014). Sociologists Adia Harvey Wingfield and Renée Skeete Alston elaborate on the work involved in undertaking racial tasks:

> Gendered organizations entrench assumed divisions between men and women *vis-à vis* constructions of work, [while] racial tasks push workers of color to display cultural practices, behaviors, and

Figure 3. "We Agree" Advertisement Campaign from Chevron Oil Company.

attitudes in response to race-related situations in an effort to prove their commonality with Whites in the organization. In contrast to the way gender inequality is perpetuated, racial inequality is maintained when minority workers do racial tasks that construct Whiteness as normative and standard. (2014:285)

By using women of color to attest to the corporation's fundamental goodness, these advertisements for BP and Chevron paradoxically uphold entrenched racism and sexism inside the company, while doing nothing to address social and environmental harms the industry imposes on women around the world.

It is not surprising that the oil industry has negatively impacted women. When an industry is run almost entirely by men and for men, it is unlikely to represent women's interests. The fact that few women hold powerful positions on the inside is surely a sign that women on the outside are ignored and their interests are suppressed.

Why is male domination so entrenched in this industry? A number of political scientists have examined the link between the oil industry and patriarchy. They argue that oil is found in the most sexist regions of the globe. But this does not mean that geology is destiny. Rather, the relationship between oil and patriarchy has to do with how oil wealth is distributed. According to UCLA political scientist Michael Ross (2012), elites in developing countries typically spend oil revenues in ways that suppress women's political rights (elites in oil-rich nations also suppress democracy more generally). He labels this "the oil curse."

Oil wealth suppresses women's rights in two different ways, according to Ross. First, political leaders in oil-rich countries often distribute oil proceeds to households in the form of direct payments, welfare programs, and government jobs. If these transfers from the government are high enough to sustain a household at an acceptable standard of living, women typically will devote themselves to domestic labor instead of engaging in paid work. The resulting traditional gender division of labor bolsters patriarchy and undermines women's political rights (Thistle 2006).

The second way that a country's oil wealth promotes male domination is by suppressing the growth of industries that typically employ large numbers of women. Wages are generally higher in countries rich in natural resources compared to their trading partners. Consequently, they do not develop manufacturing industries for export because they cannot compete on price; other countries cannot afford the goods they produce. (Economists call this phenomenon the "Dutch Disease," referring to the economic downturn in the Netherlands in the 1960s after natural gas was discovered there.) Ross points out that light manufacturing jobs are precisely those that employ women in developing countries. Without this employment niche, Ross argues, women have few economic opportunities aside from working in family-owned businesses, which do not promote women's empowerment either politically or economically.

In both cases, Ross argues, oil wealth perpetuates patriarchy by suppressing women's participation in the labor market. Ross writes, "Without large numbers of women participating in the economic and political life of a country, traditional patriarchal institutions will go unchallenged" (2012:130).

Political scientists Yu-Ming Liou and Paul Musgrave (2016) add complexity to Ross's argument, arguing that gender inequality is not only an incidental result of economic processes (as Ross maintains), but often an *intentional goal* of autocratic rulers. They argue that in oil-rich nations, leaders will sometimes purposively design policies to exacerbate gender inequality. In 1980 in Saudi Arabia, for example, leaders rolled back women's rights to signal their support for the clerics, on whom their own political legitimacy and stability depended. Women's covered and sequestered bodies became the (in)visible signs of the monarchy's support for the clerics, support that the kingdom depended upon to maintain their control over the country's vast oil resources. The economic costs of such "antisocial" policies are high, but elites willingly pay these costs to avoid alienating the conservative members of their ruling coalitions.

The view that women are pawns used by oilmen to consolidate their economic and political power is echoed by journalist Ellen Wald (2018) in her history of Saudi Aramco, the national oil company of Saudi Arabia and, at the time her book was published, the most lucrative firm in the world. She writes, "Nothing illustrates the conflict between profit and power and modern and traditional in Saudi society like the issue of women" (164). She portrays women as *victims* of the oil industry, covered up by abayas, headscarves, and laws of guardianship. The voices of Saudi women are notably absent from her book; in fact, the only Saudi women interviewed are exiles who escaped to the West to pursue their dreams of liberation. As in much of this literature, her account ignores the rich history of Middle Eastern women's resistance to their subordination (Charrad 2009).

Overall, this literature on the "oil curse" considers sexism a feature of autocratic societies, not enlightened democracies. This is a quintessential example of "orientalism," a way of seeing that exaggerates and distorts differences between Arab peoples and the denizens of the so-called "civilized" West. Europeans and Americans practice orientalism when defining Arab culture as exotic, backward, and violent, and boost their own views of themselves as superior in every way. This entrenched view dominates Western ideas and, because it does not allow

Figure 4. CEOs of the US largest oil companies testifying before the US Congress. From right to left: Lamar McKay (British Petroleum), Marvin E. Odum (Shell), James J. Mulva (ConocoPhillips), John S. Watson (Chevron), and Rex W. Tillerson (Exxon/Mobil).

those from the so-called "orient" to represent themselves, prevents accurate understanding of both groups.

Orientalism prevents Westerners from seeing how sexism pervades the oil industry in the United States and Europe. Women are not only oppressed in the Arab world; they are subjected to male domination in the West. Figure 4 is a photograph from the *New York Times,* showing oil industry executives from the five largest US producers testifying before US Congress in 2010, in the wake of the BP Gulf of Mexico oil spill disaster.

The male domination of the global oil industry is obvious in this representation of these powerful (and culpable) leaders. They occupy the top of what sociologist Raewyn Connell (1998:7) calls "the world gender order": "The world gender order is unquestionably patriarchal, in the sense that it privileges men over women. There is a patriarchal dividend for men arising from unequal wages, unequal labor force participation, and a highly unequal structure of ownership, as well as cultural and sexual privileging" (7). The male leadership of the global oil industry is a principal source of "hegemonic masculinity"—Connell's term for the dominant social practices and beliefs that embody, organize, and legitimate men's domination in the gender order as a whole. In the United States, this oil industry elite consists almost exclusively of white men, while in other oil-rich countries, the industry is headed by men drawn from the elite ethnic strata in their societies. The increasingly unregulated power of multinational oil corporations is wielded by this exclusive fraternity.

The male domination of the oil industry is reflected in and reinforced by existing scholarship. The history of oil is unabashedly a history of oilmen. Only two women are named in Daniel Yergin's 877-page Pulitzer Prize–winning account of the industry, *The Prize: The Epic Quest for Oil, Money and Power* (1991). Both women were journalists: Ida Tarbell, described as "America's first great woman journalist," who exposed the corruption in Rockefeller's Standard Oil; and Wanda Jablonski, editor of the trade publication *Petroleum Intelligence Weekly,* who worked behind the scenes to form the alliance that would become OPEC. The only other women appearing in the massive tome are consumers who are photographed either enjoying their gas-guzzling convertibles in the California sunshine, or enduring long gas lines during the oil crisis in the 1970s.[4]

Victims, journalists, or consumers. Are women really so peripheral to the largest, most lucrative, most politically powerful industry on the planet?

There are reasons to be skeptical of this depiction. Political theorist Cynthia Enloe contends that male domination depends on the systematic silencing of women's voices. In her aptly titled book *Seriously!* (2014), she points out that political and economic history is almost always written from the viewpoint of men who actively suppress women, ignore them, or refuse to take them seriously. She writes,

> The unquestioned presumptions about what and who deserves to be rewarded with the accolade of "serious" is one of the pillars of modern patriarchy. That is, being taken seriously is a status that every day, in routine relationships, offers the chance for masculinity to be privileged and for anything associated with femininity to be ranked as lesser, as inconsequential, as dependent, or as beyond the pale. (10)

This book is an effort to take women seriously, to see them as actors in this industry and not just its beneficiaries or victims. The women geoscientists in this book are contributors to the oil industry. They are among the workers who extract oil and gas from the earth. Their insider accounts reveal how male domination of this industry truly works, and what is at stake in attempting to challenge it.

TAKING WOMEN SERIOUSLY

In the academic world, it seems that the only people who take women seriously are other women. In 2007, two women writing for the Geological Society of London published the first scholarly book on women in the geological sciences (Burek and Higgs 2007). Their book reviews the history of women's contributions to the field from the 1700s to the present. The editors describe the barriers to women in the past: women were not allowed to conduct fieldwork unless a male relative was present; they were excluded from college-level teaching; and those who managed to secure professional positions were fired from their paid jobs when they married (but were allowed to stay on as volunteers). A few notable women managed to prevail and make important contributions, but the legacy of this discrimination is still evident today: they point out that only 20 women were employed as geoscience professors in the United Kingdom in 2007, a number that was down from the peak of 25 in 2004, when women held 7 percent of such positions.

The first book about women geoscientists in the United States was also a collaborative effort by women scientists (Holmes, OConnell, and Dutt 2015). Their book highlights the informal barriers to women geologists in academic careers. This volume was the outcome of the National Science Foundation's ADVANCE program, a multiyear effort to increase the number of women faculty in STEM disciplines. Started in 2001, this federal program distributed $270 million to academic research teams around the United States, who received funding to develop discipline- and university-specific strategies to overcome institutional barriers to women in the sciences. The Trump Administration put this program on hold in 2018, perhaps to signal to his conservative base that advancing women in the sciences was no longer a national priority.

It cannot be a coincidence that the two scholarly works on women geoscientists were written by women. Not that all women take women seriously: Some women feel compelled to deny or downplay gender issues as a survival strategy in a male-dominated world. Ironically, one of the first-ever articles about a women geoscientist published in the AAPG newsletter, a feature story on the career of Susan Eaton, was titled "Don't Call Her a 'Woman Geoscientist'" (Friedman 2016). Eaton, who runs her own consultancy in Canada, is quoted at the beginning of the article: "I'm tired of talking about women in the geosciences"

Figure 5. Crude Oil Prices (price per barrel in constant US dollars), 1968–2018. Source: Macrotrends LLC.

(8). By relying on a prominent woman to dismiss the importance of gender, the article unwittingly reinforces the silencing of women and the male domination of the industry.

The oil industry, for its part, started taking women seriously in the early 2000s. This is when PROWESS was formed. It was also a time when both oil prices and the number of women graduates in the geosciences were increasing (see figure 5). The industry turned to women to fill its ranks.

As figure 5 shows, oil prices rose in the early years of the twenty-first century, leading oil companies to hire more geoscientists, whose job it is to find oil and gas. After years of corporate mergers, reorganizations, and mass layoffs in the 1990s, oil companies had depleted their scientific labor force. Executives were especially worried that as senior scientists were nearing retirement, few mid-career scientists were queued up to take their places. Industry insiders referred to the impending turn-over of personnel as "the great crew change." These same executives were responsible for laying off workers during the previous downturns and mergers, so this was a problem of their own making, but a source of worry nonetheless.

The great crew change remains a concern for the industry. In 2016, Carolyn Wilson of the American Geosciences Institute estimated that "over the next decade, 48 percent of the workforce will be at or near retirement" (2016:ii), and predicted a shortage of 90,000 geoscientists (out of a total workforce of 324,000). This predicted shortage reflects a wider discourse of a STEM worker shortage in the United States. For decades, the US government, educators, and corporations have pushed for more graduates in science and engineering to feed the growing demand for knowledge workers to keep the US economy competitive (National Academies 2010; Teitelbaum 2014).

There are reasons to be skeptical of this shortage narrative. Some argue that it is actually a myth concocted by corporations to support their constant downsizing and quest for cheaper labor (Hira 2010; Teitelbaum 2014). Demographer Michael Teitelbaum argues that no objective labor analysis has ever documented "any convincing empirical evidence to confirm the existence of such generalized shortages" (2014:116). Instead, he attributes the widespread belief in the shortage of STEM workers to wildly successful lobbying and public relations efforts by corporate employers.

The business community supports the shortage narrative because it helps to garner public support for STEM education, enabling companies to lay off workers with impunity knowing that replacement workers will be readily available to backfill any vacancies that develop. Plus, brand new college graduates are significantly cheaper to hire than older, experienced workers, making it a profitable strategy to churn the workforce. Companies also claim they face a labor shortage in order to promote hiring temporary foreign workers. When tech industry executives lobby Congress to grant more H-1B visas to foreign workers, the STEM shortage may be the rationale they cite in public, but their covert agenda is wage-cutting and outsourcing (Hiltzik 2015).

In the geosciences, the supply of new graduates fluctuates over time. Some universities downsize their geology programs in response to low oil industry demand. The number of geoscience graduates plummeted in the mid-1980s in response to the precipitous decline in oil prices at the time, and has not fully recovered (Wilson 2016:48).

On the other hand, the proportion of women graduating with degrees in the geosciences has been steadily rising. Women's participation in master's programs (the beginning-level credential in the major oil companies) grew from 15 percent of students in 1975 to a peak of 45 percent in 2006. In 2015, women were about 40 percent of the students enrolled in master's programs in the geosciences (Wilson 2016:46). Recruiting women became a priority as the industry faced down the great crew change.

Companies rebranded themselves in the 2000s to convince women to join the oil and gas industry. Among their efforts, they featured pictures of women on their recruiting materials and annual reports (see figure 6).

In 2010, University of Texas sociologist Kristine Kilanski investigated the on-campus recruiting practices of oil companies. She wondered how this old boys' club was marketing itself as a viable career choice for women geoscientists after a century of ignoring them. At some of the recruiting events she attended, companies sent teams of white men to pitch their diversity

Figure 6. Recruiting materials for major oil companies.

programs. Others sent women and minority men, but they were either token members of their geoscience divisions, or employees of the personnel departments of the companies they represented. Kilanski (2011) was skeptical of these rebranding efforts.

Ultimately, however, it did not matter how well or how poorly the recruiters performed. The starting salaries the industry offered were two times the amounts paid by alternative employers (such as government, environmental consulting, or academia). A salary survey by the American Geosciences Institute found that *every* new graduate earning over $90,000 per year was employed by the oil and gas industry (Wilson 2016:4). The industry also offers graduate students the advantages of paid internships, scholarships, and signing bonuses.

Because many college students today graduate heavily in debt, these high oil industry salaries hold obvious appeal. However, as a group, geoscience majors tend to be environmentalists. Many are drawn to the earth sciences out of a deep interest in nature and ecology. For them, the oil industry can be a hard sell. The industry manages to sidestep the hardcore "dark green" environmentalists by targeting their recruitment at so-called "oily schools," universities in places like Oklahoma, Texas, and Colorado, with well-entrenched petroleum programs subsidized by the oil industry itself. On my campus, for example, the geosciences are located in the Jackson School, named after an oilman-turned-philanthropist, while the geoscience program at the University of Oklahoma is even more brazenly titled the "ConocoPhillips School."

Although there are no reliable statistics on corporate hiring practices, industry reports claimed that during the upturn in oil prices, the number of women hired was proportional to their representation among new graduates, at least in the larger companies (this was the case at GOG, the company I studied). But the members of PROWESS were nevertheless worried. Their experience taught them that women would enter the industry, but not stay long.

Thus, with the support of PROWESS, in 2012 I embarked on a longitudinal study with Chandra Müller to follow a beginning cohort of scientists over a five-year period. Because GOG had strong ties to PROWESS, managers there were willing to provide us an email list of the 360 scientists and engineers they had hired in the previous five years. We annually surveyed this group of mostly young (25-to-35-year-old), mostly white men and women about their experiences working in the industry. Our goal was to document the stumbling blocks they encountered in their early career development. We thought that if we could identify any barriers associated with gender, we could help the company design programs that would staunch the flow of women out of these STEM jobs.

Because of the design and timing of the study, we were able to follow this cohort over a period of boom and bust. I interviewed a subsample of 44 of them over a three-year period, at the height of the oil boom in 2014, and also at its nadir in 2016. I talked to men and women excited about beginning their careers at GOG, and later struggling to keep their jobs. Our survey revealed that women bore the brunt of the cuts, with half of the women compared to a third of the men losing their jobs (Bosky, Muller, and Williams 2017). My in-depth interviews with a subsample of them allowed me insight into the personal impact of corporate downsizing on scientists and engineers as it was unfolding around them. These interviews also shed light on how the industry was able to renege on its commitment to diversity without suffering any penalty or even pushback from its STEM workforce. The result is a cautionary tale about gender and racial inequality in the new economy.

THEORY OF GENDERED AND RACIALIZED ORGANIZATIONS

Throughout this book, I refer to the company that I studied as a "gendered" and "racialized" organization (Acker 1990; Britton and Logan 2008; Gorman and Mossari 2019; Ray 2019; Williams, Muller, and Kilanski 2012; Wingfield and Alston 2014). This conceptual framework highlights how workplaces are set up to advantage certain groups—white men in this instance—and discriminate against others. These disparities are evident at every stage of a professional career, from recruitment, to promotions, to transfers, to layoffs and firings. I draw special attention to how certain innovative features of work organizations in the new economy—including the implementation of career maps, networking, and teamwork structures—reward elite white men and qualities associated with hegemonic masculinity, while establishing obstacles for other groups.

In the United States, it is illegal to discriminate in employment based on gender and race/ethnicity. This is in contrast to the not-so-distant past, when work rules in virtually all employment sectors openly discriminated against women from all racial and ethnic backgrounds, as well as against racial/ethnic minority men. It was legal and, in many workplaces, perfectly acceptable to segregate the workforce by race and gender, with the best jobs reserved for white men. Prior to the Civil Rights Movement, "Help Wanted" ads reflected this taken-for-granted segregation by listing jobs separately for men and for women. In these job listings, employers sometimes specified the desired race and ethnicity of job applicants as well. In the oil and gas industry, the legacy of this legal segregation lives on in the titles of some jobs, including "land man" and "company man," and housing for workers at well sites, which are still called "man camps."[5]

In the 1960s, it became illegal to discriminate in employment and to segregate workers based on gender and race. Work rules were rewritten throughout the labor market to conform to these new requirements (Dobbin 2009). Many of the white men employed by the oil industry resisted these changes, in some instances sending their wives to picket lines to protest the gender- and racial-integration of their jobs, but the battle for legal de jure segregation was lost (Priest and Botson 2012).

As a result of the Civil Rights Movement, organizations today have gender- and race-neutral work rules, yet white men still dominate the top jobs. Theories of gendered and racialized organizations attempt to explain why. They posit that organizational rules remain biased, even if no longer explicitly so, because gender and race stereotypes are embedded in organizational practices. All personnel decisions are filtered through a lens of gender and race. At the point of hiring, employers select workers for gender- and race-typed jobs, with people of color concentrated in the lowest rungs of the organization (Wingfield and Alston 2014; Ray 2019). When deciding on criteria for promotion to management, employers will typically choose candidates who exhibit authority, ambition, and confidence—qualities closely associated with elite masculinity. These criteria favor white men for promotion opportunities, even though they are not the only people who possess these qualities. Discrimination occurs because employers rarely recognize these traits when expressed by women or nonwhite men, or if they do, they stigmatize these employees for violating norms about deference in the workplace. For instance, white women and minority women might be criticized for not being "feminine" enough if they demonstrate authority and ambition, while minority men exhibiting these qualities may be feared and not respected (Wingfield 2009, 2010). This is just one source of discrimination that may exist in ostensibly "equal employment opportunity" organizations.

Employers often claim they are innocent, arguing that they are not even aware when they engage in such discriminatory practices. Doing so may enable them to escape legal culpability, as the antidiscrimination law specifically prohibits only *intentional* bias in the workplace (Lee and Ahmad 2019). In support of this narrative, a cottage industry of consultants developed to uncover and correct the "unconscious biases" of managers (Berrey 2015; Dobbin 2009). In so-called diversity training sessions, these consultants teach managers to recognize their prejudices on the assumption that, once enlightened, their vetting practices will naturally adjust, and fairer outcomes will be achieved.[6]

Sociologists agree that the mechanisms that produce social inequality in the workplace are sometimes invisible, but their focus is more on organizational, and less on individual, forms of bias. A plethora of "glass" metaphors describe these unseen structural forces, including the glass ceiling, the glass escalator, the glass cliff, and the glass slipper. These terms designate the ways that organizations advantage some groups at the expense of others, especially women and minority men. As organizational scholar Karen Lee Ashcraft (2013:15) writes: "The utility of glass metaphors lies in their capacity to name and evoke systematic patterns that are otherwise elusive. They provide tangible abbreviations or proxies that redirect us from individual explanations (e.g., willful prejudice) to institutional accounts, surfacing hidden dynamics at work that call for further exploration." The invisible processes that result in social inequality at work are institutionalized in the oil and gas industry, meaning that they are taken-for-granted as normal and rational business practices, but they ensure that white men rise to the top. Once made visible, however, they can be shattered, which is the whole point of projects like this one.

That said, not all inequality-producing mechanisms today are unconscious or invisible. Some employers intentionally discriminate based on gender, race, and other statuses when making employment decisions. Several respondents in this study told me about managers and supervisors who directly and intentionally discriminated against them. Layoffs are not invisible, either. Layoffs present an opportunity to exercise managerial bias (Byron 2010), yet this increasingly common experience has yet to be incorporated into theories of gendered and racialized organizations. *Gaslighted* adds downsizing to the long list of organizational practices that reproduce social inequality in the workplace. As the US oil and gas industry shows, when laying off workers becomes a routine and legitimate business practice, it can guarantee the continued elite white male domination of corporate America.

ORGANIZATION OF THE BOOK

The book focuses on the career narratives of STEM professionals who work for a giant oil company. I begin by describing their pathways into the oil and gas industry. Bringing more women and minority men into STEM careers has become a national priority. Chapter 2 starts by examining the motivations to become scientists. I asked my respondents about their youthful exposure to science and their earliest memories about wanting to become a geologist. I then explore their college experiences and their decision to work for an oil and gas company. Typical geoscience majors are environmentalists, drawn to their subject out of a love of the outdoors and a desire to preserve and protect the earth and its natural resources. Many of those I interviewed considered the oil industry "evil" during their college days. The industry succeeded in luring them in with the promise of access to exciting new technologies, mountains of scientific data, and very large paychecks, causing them to either reorder their priorities or reevaluate their negative stereotypes about the industry.

While white women have made inroads into the geosciences in recent years, this is not the case for racial/ethnic minority women or men. Geology is among the whitest of the natural sciences. I explore the history of racism in the discipline, and suggest both causes and potential solutions to the lack of racial diversity in the college-to-industry pipeline. The whiteness of the college major is reflected in industry demographics—where most of the scientists who are categorized in US companies as "people of color" are foreign-born—but it is exacerbated by corporations that target their recruiting efforts at predominantly white colleges and universities.

Subsequent chapters delve deeply into the career narratives of the 14 women and 8 men geoscientists I interviewed over time at GOG. I divide my respondents into three groups depending on how their careers unfolded over the course of the study: chapter 3 focuses on those who remained at GOG throughout the entire study; chapter 4 spotlights those who voluntarily left the company to work at another oil company; and chapter 5 presents the accounts of those who left the oil industry altogether. By focusing on their individual experiences, I hope to convey the unique pathways of those I followed over time, while putting their individual narratives in organizational context. Doing so, I am attempting to exercise my "sociological imagination"—C. Wright Mills's (1959) term for situating personal experiences within the context of global capitalism. I highlight their individual agency as they navigate the constraints of the corporate world.

In chapter 3, on the "Stayers," I focus on the career narratives of the three (out of the 14) women geoscientists in my study who remained at the company after surviving four rounds of layoffs at GOG. (Four men are also in this group.) I explore what these "successful" scientists share in common and what lessons their stories contain for others who may want to follow their path.

Perhaps surprisingly, this is not an upbeat chapter. Unlike many other accounts of successful women scientists, the three women I highlight in this chapter do not tell heroic stories of conquering sexism through passionate commitment to their science. Instead, theirs are survivor stories filled with compromises they made to keep their jobs during the industry downturn. To survive a sexist work environment, these women had two choices: either support the company wholeheartedly, or keep their heads down and their mouths shut. The constant threat of layoffs meant that any expression of dissent could put them on the chopping block. I argue that silencing geoscientists has dire implications for both diversity in the corporate world and protection of the environment.

Chapter 4 looks at the experiences of those who voluntarily left GOG to join a different oil company. They represent the kind of employee attrition that companies wish to stop. In fact, GOG agreed to participate in this study during an oil boom, when concerns about voluntary attrition were at a peak. At the time, the geoscientists in my study were fielding calls from professional recruiters every week, and GOG wanted to know what they could do to stop them from leaving.

The eight people (five women and three men) in my interview sample who left GOG for another oil company did so for different reasons, depending on whether they left before or after the industry downturn. Those who left "before" did so for mostly positive reasons—they perceived better personal or professional opportunities at another firm. They had only good things to say about GOG, even though ultimately they felt that the company was not the best fit for them. In contrast, those who left "after" the recovery started did so to escape a toxic work environment. Multiple rounds of layoffs had resulted in intolerable working conditions in the company, and they leapt at the first opportunity to leave. Their stories provide a cautionary tale to employers: stemming voluntary attrition during good times may require companies to stop treating workers as disposable when times are tough.

I turn my attention to the five women who left the industry altogether in chapter 5 (there are no men in this group in my sample). They are the victims of industry downsizing. One was subjected to "forced retirement," a company policy that targeted all workers over age 50 for layoffs. The others were all mothers of young children, exposing the particularly brutal form of GOG's "motherhood penalty" (Correll, Benard, and Paik 2007).

These personal narratives delve deeply into the traumatic experience of layoffs. Yet their stories not only expose the human suffering behind the numbers; they also belie their collective exposure to gaslighting. They reveal how a particular multinational corporation is structured in ways that favor the men who dominate their industry. The rhetoric of diversity may have obscured these preferences during good times, but when times got tough, management's decisions about whom to lay off revealed their built-in favoritism for white men.

Extending organization theory, this chapter highlights how teamwork, career maps, and networking contribute to women's vulnerability to layoffs. Nominally considered benign gender- and race-neutral organizational processes, these three features of work in the new economy introduce bias into how workers are evaluated, the opportunities they receive, and their chances of being laid off.

In chapter 6, I circle back to gaslighting. I urge scholars to center the experience of downsizing in their analysis of workplace inequality. Employers in the United States have almost unlimited power to fire workers through the mechanism of layoffs. Even once-protected professionals face precarious working conditions today. As the oil and gas industry shows, racist and sexist biases determine who gets in, who gets to stay, and who must leave. Like canaries in the coalmine, women scientists in the oil industry offer an ominous warning about the future of equality in corporate America.

CHAPTER TWO

The Oil and Gas Pipeline

We Need More Women in STEM!

The shortage of STEM professionals alarms US policy makers. The National Research Council's ominous report "Rising above the Storm" (National Academies 2010) warned that maintaining American global dominance requires drastic steps to develop new STEM talent. The supply of STEM workers currently falls far short of demand, the report claimed. As a result, the United States is trailing other countries in developing technical solutions to major global challenges. Getting more women to pursue careers in science and engineering can fill this pressing need, according to the report. Because men greatly outnumber women in these fields, women constitute an untapped reservoir of STEM talent. More women in STEM will make America more economically competitive, bring new perspectives into the scientific establishment, drive more innovation, and promote gender equality in society (Hira 2010; Metcalf 2010).

Or so the story goes.

Just how real is the STEM shortage? Some researchers insist that there is no shortage of STEM workers in the United States at all. More than enough qualified people exist to fill available positions (Hira 2010; Teitelbaum 2014; Metcalf 2010). This is certainly evident in the academy, where the supply of new PhDs greatly exceeds the demand for full-time faculty. However, even in industry, there are more STEM workers than there are jobs for them, and this has been the case for some time.[1] Nevertheless, a broad consensus exists among political elites about a dire shortage of engineers and scientists, despite scant empirical evidence of demand outstripping the supply of scientists and engineers in the United States (Teitelbaum 2014).

So why the ominous warnings of a STEM shortage? One reason is because many groups—including employers, educators, and immigration lawyers—stand to benefit from increasing the supply of scientists and engineers. Employers are especially eager to increase the number of STEM workers because doing so would increase competition for jobs so they can lower wages (Teitelbaum 2014). Increasing the number of scientists and engineers could also enable employers to replace their seasoned employees with cheaper and assumedly more compliant (i.e., younger, female, immigrant) workers—a much different gloss on why employers might target women for STEM jobs. Replacing older employees with new graduates also guarantees employers' access to the most up-to-date skills. Because STEM education is expensive, employers may prefer not to invest in the retraining of their existing employees. Hiring young people fresh from college guarantees a labor force familiar with the latest technologies and methods, effectively outsourcing training costs to universities.

For advocates of women in STEM, however, the goal is not to help employers to lower wages or churn the labor force, but to undermine sexist stereotypes. Even today, many believe that women lack the innate capacity to do science, a view that perpetuates gender inequality in society as a whole. The stereotype that boys are naturally better at math and science is widespread, including among high school math teachers (Riegle-Crumb and Humphries 2012; see also Correll 2001; Hill et al. 2017). Larry Summers famously expressed this view when he was president of Harvard University; more recently, the Google engineer James Damore stated the same opinion (Thébaud and Charles 2018). Notably, both men left their jobs soon after they publicly endorsed these sexist stereotypes.

These stereotypes persist despite the fact that girls outperform boys in high school math, which is the gateway to science careers. Today, girls are just as likely as boys are to take and excel at advanced math classes in high school, and they make up almost half of all math majors in college. Physics is an exception, but even in that subject, half of all high schools have achieved gender equality or even a small female advantage (Riegle-Crumb and Moore 2014). Nevertheless, the stereotypes persist that girls are either not interested in or not qualified for STEM careers. Challenging those stereotypes and encouraging more girls to pursue these subjects have become a key issue for feminists fighting for gender equality.

I should mention that not all feminists agree that boosting the number of women and girls in science is a worthwhile political goal. Although women may be perfectly capable of excelling in STEM disciplines, some feminists argue that there are good reasons why women might want to avoid careers in science and engineering. Following the wisdom of poet and writer Audre Lorde (1984), who famously argued that the "master's tools" will never dismantle "the master's house," these feminists question the view that women should pursue equality by emulating men. Sarah Giordano (2017), for example, left a career as a neuroscientist to pursue feminist science studies. She does not believe that sciences are inherently superior to all other disciplines. She argues that science is not more objective, systematic, or valid than work done in the humanities, which is too often dismissed as subjective and "soft." This gendered binary is inaccurate and it supports male domination by devaluing qualities associated with women and femininity. Likewise, social theorist Patricia Hill Collins (1999) observes that the same dichotomous thinking that shapes race and gender also seems essential to Western science, which defines itself as abstract, universal, and rational—all qualities conventionally and exclusively associated with white men (see also Johnson 2007). Because STEM fields promote hegemonic masculinity and white supremacy, women may be better off rejecting—or "escaping"—these disciplines and developing alternative ways of knowing (Giordano 2017:11; see also Subramaniam 2009).

Such views rarely get a hearing in the US policy world, where the dominant narrative insists that our economy is suffering from a STEM shortage and it is in everyone's best interest to get more women and girls to pursue careers in these fields. To this end, scores of nonprofits sponsor programs to encourage girls' interests in science and engineering. One recent example is the Ad Council's "She Can STEM" project (see figure 7). This public service advertising campaign encourages girls to consider STEM careers by promoting images of stylish women role models on the frontiers of science. According to the website, this campaign is funded by "five of the world's biggest tech brands," including Microsoft, Google, and IBM (all have laid off STEM workers in recent years).

In addition to nonprofits, companies are getting into the act. Toy makers like Goldie Blox and Legos market products to girls to make engineering and computer science more appealing to

Figure 7. Screen capture from the Ad Council's "She Can STEM" campaign, 2019.

them.[2] Meanwhile the US Congress passed bipartisan legislation to enhance STEM education in schools with a focus on girls. Universities pump money into STEM programs and vigorously recruit women students into these majors, while scores of federally funded researchers study how to keep them there.

One of the biggest federal initiatives to increase the number of women in STEM is the ADVANCE program of the National Science Foundation. An outgrowth of an infamous study of gender discrimination at MIT that documented gender inequality in the distribution of resources among academic scientists, the goal of the ADVANCE program is "to increase the representation and advancement of women in academic science and engineering careers, thereby contributing to the development of a more diverse science and engineering workforce." NSF-ADVANCE has contributed over $270 million to support social science research and to transform higher education (NSF 2017).

Research funded by this federal program often uses the metaphor of a "pipeline" to describe and diagnose the barriers that women students face once they enter college.[3] Owing to institutional barriers and the "chilly climate" they confront in universities, many of the women who choose STEM majors "leak" out of the pipeline each year, so that by the time of graduation, only a small number of qualified women are available to enter graduate programs or begin careers in STEM. Research studies funded by NSF-ADVANCE aim to transform university education to make it more accommodating and welcoming to women, and to encourage more women to enter the professoriate to serve as role models for future generations of women scientists and engineers.

THE GEOSCIENCE PIPELINE

This chapter investigates the pipeline for women in the geosciences. Drawing on my interviews with men and women working for GOG, I ask how they got interested in the discipline, their experiences in college, and their decision to go to work for the oil industry. My goal in this chapter is to identify the ways in which the pathways into the discipline and into the industry are gendered and racialized. Women make up about 40 percent of geoscience majors in college, making this one of the most gender-equitable of all STEM disciplines in the United States. However, most of these women students, like most of the men students, are white. The geosciences are the least racially diverse of all the majors in the natural sciences (Bernard and Cooperdock 2018; NCSES 2019). In this chapter, I address the whiteness of the discipline, and ask why the geosciences fail to attract minority women and men even as they succeed in attracting white women and men. I asked the scientists I interviewed to recollect what they found appealing about the geosciences and how they decided to major in the discipline. By examining these scientists' narratives, I reveal discourses and practices that sustain hegemonic masculinity in the university context, and that perpetuate the whiteness of the discipline.

Following graduation, about two-thirds of geoscience graduates find employment in the oil and gas industry, with women represented at about the same proportion of new hires as their share among graduates. In this chapter, I also address this transition-to-work: How do geoscience students connect with the industry and decide to take jobs there? What are the costs and benefits of that choice, from their points of view?

Companies play an active role in recruiting geoscience graduates into the oil and gas industry. This is sometimes a hard sell, as many potential recruits consider themselves environmentalists who never imagined working for the "evil" oil industry. In the interviews, I learned about the extensive effort that GOG put into convincing some of these scientist-environmentalists to work for them. These recruiting stories reveal another set of gendered and racialized mechanisms that more or less guarantee the continued whiteness and male-domination of the oil and gas industry.

Throughout my discussion, it is important to bear in mind that I am discussing the experiences of men and women employed by one specific oil and gas company. My respondents are not a cross section of geoscience graduates, nor are they a cross section of geoscientists in the oil industry. Instead, they represent a group of workers vetted and hired by one company in a competitive industry. Their accounts provide insight into the processes that led to successful graduation outcomes and careers for one particular group, but leave out the experiences of those who took other paths.

In the interviews, I asked respondents to recollect how they came to work in their chosen field. This is not an easy question for anyone to answer. In response to such questions, people usually tell stories. These stories are not necessarily accurate descriptions of the past (as if such a thing were even possible), but they do convey information about what respondents value and how they want to be perceived. In-depth interviews, like all social interactions, are occasions for what sociologists sometimes refer to as the "presentation of self." As such, they offer opportunities for respondents to show themselves in a favorable light, communicating what they regard as their good qualities. In their responses to my questions, they could represent themselves as honorable, deliberative, or adventurous, for example. These performances are gendered as well. The interviews reveal how they enact and negotiate society's prescribed rules for how men and women should behave, and what feelings are deemed appropriate to express.

In addition to sharing information about themselves as individuals, the stories my respondents told give insight into their disciplinary training. Just as nurses learn to express a desire to "help people" and sociologists are taught to express concern about social justice, geologists learn to profess the specific values promoted by their discipline. I examine these accounts for what they can reveal about the respondents' idealized image of a professional geoscientist.

Finally, the interviews reveal something about the culture of the labor market. Since these respondents were all successful in their job search, I can infer from their accounts not only what they value and their discipline values, but also what the company that hired them values. From their stories, I learn what GOG considers an ideal scientific worker in the oil and gas industry, and how gender and race are embedded in that ideal.

LOVE OF NATURE: MAJORING IN GEOLOGY

One of the first things I learned about geoscientists is they are outdoorsy people. This is the standard way the discipline represents itself. Photos of people enjoying the wilderness fill the university webpages of the major geoscience programs. Backpacks, hiking boots, and camping equipment are common sights even in faculty portraits. Likewise, conference brochures are illustrated with stunning pictures of the natural environment, and offer attendees tours of nearby national or state parks.

Not surprisingly, when asked to look back on their decision to major in geology, many of my respondents professed a love of nature and the environment, a narrative that shows up in other studies of geoscience majors as well (Huntoon and Lane 2007; Stokes, Levine, and Flessa 2015). Some told me that their love of nature was cultivated during family vacations to national parks. One geoscientist recalled hiking in the woods with her father, who inspired her inquisitiveness about nature. Another talked about growing up on the beach, which naturally stimulated his love for the physical world. Others told me that college kindled (or rekindled) their love of nature. The following are quotes from three women geoscientists:

> I heard geology was the easiest thing you could take to fulfill the requirement. So I said, yeah, I'll do that. But I took the first class and thought, I love this! I've thought about rocks since I was a little kid. I was obsessed with collecting rocks. I tried to sell them on the side of the road.

I happened to take this geology class because I needed a science elective. I just fell in love with it. Everything about it: the outdoors aspect, the natural aspect, seeing things three dimensionally in my head, which I think feeds on that artistic side. I honestly had no exposure to it as a child whatsoever. Arkansas is not exactly fond of the earth sciences in elementary school, middle school, or high school (laughs). So, yeah, my first exposure to it was in college.

I started out an education major. Took a few geo classes. Found out it was mandatory to go to the Grand Canyon, and found out that I loved geology and I was not cut out to be a teacher, so I switched.

Almost all of the geoscientists I interviewed, both men and women, attested to a love of nature. The ubiquity of this motif marks it as a discourse—an appropriate disciplinary value that geoscientists are supposed to embrace. A good geoscientist is someone who loves nature.

This talk of love struck me as strange from my outsider perspective. It would be odd for a social scientist to profess a love of society. The engineers I interviewed did not talk about love, either, describing their motivations for selecting their major in much more instrumental terms. Employability and career flexibility were important considerations for the engineers, not passion or love. One engineer told me he simply picked "the highest paying major." In some cases, the engineers' parents who were funding their educations made the decision for them (several were children of engineers; see Jacobs, Ahmad, and Sax 2017).

To be clear, some engineers love the work they do, but with rare exception, they did not talk about picking their college major in these terms. I suspect that it might be taboo to do so. Just as "love of nature" is a discourse, so are the instrumental values and family influences highlighted in the engineers' narratives.

Other scholars have noted the penchant among natural scientists to express love for the objects they study. According to Giordano (2017), this is especially common in the narratives of the first women allowed to pursue scientific careers. Consistent with this, passion and determination are the dominant themes highlighted in the volume *Anomalies: Pioneering Women in Petroleum Geology, 1917–2017* (Gries 2017), a publication distributed by the American Association of Petroleum Geologists.

Campaigns to encourage girls in STEM disciplines, including engineering, treat passion as a prerequisite for success. For instance, the website of "She Can STEM" features this advice from Danielle Merfeld, Chief Technology Officer at GE: "Do what you love, and you'll always be successful." The Society for Women Engineers sponsors a program for middle school students that helps girls "draw connections between their career passions and engineering." And Google's program to interest young women in STEM careers claims: "Girls start out with a love of science and technology, but lose it somewhere along the way. Let's help encourage that passion in teen girls."[4] Even sociologists contribute to the passion discourse. In an article examining barriers to women in STEM careers, sociologists Sarah Thébaud and Maria Charles (2018:12) conclude that girls should be taught to become "just as passionate about computer science and engineering as they are about teaching and nursing."

An advertisement for StatOil emphasizes the passionate commitment of a geoscientist named Allison, shown asking the question, "Where did they all go? I mean ... literally" (see figure 8). The accompanying text continues:

> Passion goes a long way.... To know where to search, our passionate geoscientists develop advanced technology to visualize what

Figure 8. StatOil Advertisement.

the planet looked like, back when it was all one continent.... Thanks to our passion, StatOil has become one of the most successful exploration companies in the world today.

Surprisingly, women in STEM are characterized as passionate despite the experience of discrimination. An *Atlantic* magazine article on women engineers titled "Why Is the Silicon Valley so Awful to Women?" begins, "The dozens of women I interviewed for this article love working in tech. They love the problem-solving, the camaraderie, the opportunity for swift advancement and high salaries, the fun of working with the technology itself." This love discourse is so common that I started searching for accounts by women in STEM who are not passionate about their work. Does any woman scientist ever experience boredom, indifference, or antipathy about their jobs? Sociologist Dana Britton (2017) found that articulating these sentiments is uncommon and perhaps taboo, even among women scientists who work in demonstrably "chilly climates."

I suspect that all this talk about love also may be a way for women to legitimize their presence in an otherwise hostile environment. Attesting to childhood passions proves that at least some women have innate proclivities for STEM. By representing themselves as driven, without regard for barriers that society imposes on them, successful women portray themselves as rising above prejudice and discrimination. In this romantic discourse, love conquers all.

This discourse can paradoxically exclude women, however. The idea that passion drives successful women scientists assumes that those who are not passionate do not apply. If few women enter these disciplines, the implication is that most women are not passionate about STEM, or else they do not love these fields enough to overcome any obstacles in their way. This discourse blames women—and not the organization or culture of science and engineering—for the male domination of these fields.

As it turns out, passion may have little to do with women's persistence in STEM. A study by Nadya Fouad and her colleagues (2016) found that interest in their chosen field did not distinguish women engineers who left from those who stayed. In their study of more than 500 women engineers, the only difference they were able to detect between these two groups is that women who left experienced lower levels of workplace social support than those who stayed. In other words, having a strong network of advocates helped to prevent women's attrition from engineering. In their study, professing enthusiasm or passion for engineering did not distinguish the stayers from the leavers.

Understanding the "love of nature and science" as a *discourse* reveals its embeddedness in fields of power, offering not only insight into the gendering of the discipline, but also a clue to the whiteness of the profession. People of color stand in an ambivalent relationship to discourses about nature. Scientific racism—the belief that white people are biologically superior to all other groups—is the foundation of modern white supremacy. Proclaiming "love of nature" or "passion for science" discounts the history of abuses of Black, Latinx, and Native American people conducted in the name of science, and the history of community resistance to these abuses. As a discourse, this particular motivation promotes whiteness in the discipline (and in the professions more generally; see Rao and Neely 2018).

Moreover, the childhood cultivation of the "love of nature" is not a common experience among people of color. Members of racial and ethnic minority groups are not frequent park visitors, a result of many factors, not least the racist history of exclusion and displacement in those spaces. Some parks are located on lands dispossessed from Native Peoples who were forcibly removed when the parks were first established. The nineteenth-century national park movement was the project of white supremacists who intentionally excluded nonwhite Americans. Not until the civil rights era did state parks in Texas even allow Black and Mexican American visitors. For some people of color today, wilderness areas evoke fears of violence and even lynching, not the contemplative enjoyment of nature (Byrne and Wolch 2009; Scott and Lee 2018; Spence 1999; Taylor 2016).

In response, some geoscience educators are developing programs to expose minority youth to positive outdoor experiences.

Their aim is to cultivate in them the "love of nature" that would then help to diversify the geosciences. Without such positive experience, the educators fear, few racial/ethnic minority members will ever feel welcome in the discipline (Bernard and Cooperdock 2018).

However, I suspect that changing the discourse might be a more effective strategy. African Americans, Latino/as, and Native Peoples have long histories of engaging the natural world, including through farming, hunting, and healing, none of which figures prominently in mainstream environmental discourse (Finney 2014). Relating the study of geology to the traditions of excluded groups may aid in diversifying the college major (see also Stokes et al. 2015). It might also help to reorient the environmental discourse away from an ecocentric focus and toward a human-centric approach. Since the Civil Rights Movement, African Americans have been on the front lines fighting for environmental justice (Bullard 2019). In stark contrast to the mainstream environmental movement, which has historically prioritized issues such as wilderness preservation and the protection of endangered species, the leaders in the environmental justice movement, many of whom are women of color, target issues such as industrial pollution in minority neighborhoods (Melosi 1995). The geosciences might become more inclusive if the discipline promoted a discourse of environmentalism linked to the health and well-being of Black, Latinx, and Indigenous people instead of the contemplative enjoyment of nature.

Historically, white women also encountered barriers to the great outdoors, but many of these have since been overcome. Wilderness areas were once the exclusive purview of white male explorers. Accounts of early women geoscientists describe struggles to overcome significant taboos about working and living outside. The first women geologists had to be chaperoned by their husbands on scientific field trips.

The white male domination of geological fieldwork is reflected in the historical icon of the intrepid fieldworker— a bearded white man. This ubiquitous image is the subject of a film and photography exhibit called the "Bearded Lady Project: Challenging the Face of Science."[5] The photography exhibit consists of individual portraits of two hundred women geoscientists wearing pasted-on facial hair. A curated collection of these images at the Smithsonian Institution's Museum of Natural History explained its intent (see figure 9): "These images harken back to the stylized photographs of male paleontologists in historical texts. The beard, a potent symbol of masculinity, is added to the women's portraits to highlight the inequities and prejudices that persist in the sciences." The "Bearded Lady Project" highlights the gendered cultural barriers between women and the practice of science. The women who posed for these portraits, like several of the geoscientists I interviewed, were determined to break down the association between roughing it in the outdoors and masculinity. But making gender trouble poses its own risks. Not everyone is willing or able to engage in acts of cultural subversion in white male–dominated spaces. Those who make "gender trouble" may not persist long in the profession.

I had the opportunity to view these photos at the Smithsonian exhibit in March 2020. On my way out of the exhibit hall, I did a double take, as the very next exhibit I encountered underscored the point by featuring an iconic geologist, a bearded white man (see figure 10).

Even today, geological field sites are a white man's world, considered hostile territories for women. Virtually all geoscience programs require students to conduct research off campus, and

Figure 9. Images from the "Bearded Lady Project" at the Smithsonian Natural History Museum, March 2020. Dr. Karen Chin, Paleoecologist, University of Colorado Boulder; Leckie Lab: Adriane Lam (PhD), Raquel Bryant (MA), Serena Dameron (PhD), students of Micropaleontology.

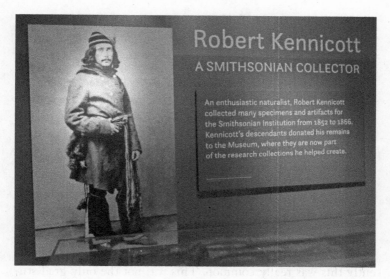

Figure 10. The iconic geologist at the Smithsonian Natural History Museum.

many require overnight stays in remote areas. Although field requirements may draw some students into the geoscience major, as was the case with several of my respondents, recent years have seen an outpouring of reports from women describing harassment and assault in the field. The sexual vulnerability of women students has been a topic of geoscience conferences, seminars, and websites.

Anthropologist Kathryn Clancy and her colleagues conducted an online survey of field scientists about their experiences of sexual harassment (Clancy et al. 2014). More than 650 women and men responded from a variety of disciplines, including anthropology, archeology, and geology, and from all different career stages, from graduate students to senior faculty. A quarter of the women said they had experienced sexual assault while working in the field, in some instances perpetrated by their professor or supervisor. One of their interviewees reported,

> The head of the site would systematically prey on women.... I was in my bed one time and he was with a married master's student and she was basically just crying and she had to leave the site because he was seducing her and she couldn't say no.... I had to serve as a kind of bodyguard for some of these women and some of them would sleep on the floor because they were afraid he was gonna come into the room at night. (Nelson et al. 2017:713)

A geologist who received her master's degree in 1985 told me a similar story. Describing a fellow graduate student in her program who worked at a remote field site, she said: "[Her advisor] came into her bedroom one night. She kicked him out, but apparently this was really common. This was not the only grad student that this had happened to. And she left. I know that was part of the reason she left the program." My respondent reported the abuse she had witnessed to the head of the department, but the offending professor did not lose his job as a consequence. He was still on the faculty when I interviewed her. Accounts like these convinced the NSF-ADVANCE program to fund the development of a sexual harassment training program specifically tailored for geologists while on field trips (ADVANCEGeo 2019).

While many survivors of sexual harassment and assault leave their professions to escape the abuse (McLaughlin, Uggen, and Blackstone 2017), others stay. After all, it is difficult to abandon a career after making a substantial investment in training, even for those who suffer this abuse. Moreover, there are no obvious career choices that are free of sexual harassment. Although sexual harassment and assault may be endemic to field sciences (Hanson and Richards 2019), these experiences are unfortunately common throughout the labor market.

In summary, my interviews reveal that the path to becoming a geoscientist is gendered and racialized. The discipline attracts students who want to pursue outdoor adventures in the wilderness—interests that some developed as children. The white masculine image of the iconic geologist might be off-putting to some women, but it attracts others intent on challenging the stereotype. The association of geology with wilderness adventure might also exclude men and women of color, who may experience these spaces as unwelcoming and even dangerous. Recognizing the trajectories into geology as gendered and racialized may help to explain the whiteness of the profession. Not that the geosciences are unique in this regard: all college majors contain assumptions about the ideal student that reflect and reinforce the gender and racial composition of the student body. In the case of the geosciences, the love of nature is the dominant discourse that brings white college students into the major, while simultaneously standing in the way of diversifying the discipline.

THE EVIL OIL INDUSTRY

How do geologists who "love nature" end up working for an oil company? Looking back, several of those I interviewed told me they never thought that their future would include working for oil and gas. Instead of polluting the environment, this woman geologist thought she would be working to clean up the environment.

> I had no intention of going into oil whatsoever.
> cw: What kind of career did you have in mind?
> I wanted to clean up ground water contamination. That was my thought while I was in grad school. I was going to go out and save the world and clean up the world.

A number of my respondents characterized themselves as counterculture people. "Petroleum never crossed my mind," another woman told me. After confirming that I would not share her responses with her employer, she admitted, "I was a hippie who thought I would never sell out and go to work for an oil company." Her dream was to get a job working outdoors, or possibly get a PhD and become a professor. Then, she said, "Reality set in":

> I really had no idea what I was stepping into when I got to graduate school. Because I was in such a small school for my undergrad. I was such a big fish in a little pond and then I get to grad school, and I realized, Holy crap! There is a lot of amazing science going on that I had no idea about. And so the lack of self-confidence kicked in: I can't do this! There is no way I can do a PhD in this. And then desperation set in: I gotta get a job! Crap! I don't want to move back to Arkansas. I gotta get a job. So I actually applied for Teach for America. I got interviewed and everything. And then I just didn't get it. When that fell through, I said, All right, I'll throw my name in the hat for some interviews with oil companies and see what happens. So that is what happened.

A sense of desperation about her career prospects propelled her to the oil industry. Because she was attending one of the top petroleum geology programs in the United States, opportunities for interviews leading to lucrative internships and jobs in the oil industry were plentiful, and they held an incessant monetary appeal.

During boom times, the oil industry aggressively recruits graduate students in the geosciences who attend "oily schools," including the University of Texas at Austin, the University of Oklahoma, and the Colorado School of Mines. These are some of the top-ranked programs in petroleum geology. They are also predominately white institutions. Students enrolled in these programs have access to summer internships, scholarships, and other enrichment activities all bankrolled by the oil industry. Students who attend graduate school at universities outside of this select group can apply directly to the companies for internships, or else apply through the AAPG's student placement program. In comparison with students at oily schools, they must expend considerably more effort and personal resources to access these opportunities, and they have lower chances of successful placement.

The fact that oily schools are predominately white is not a coincidence. School segregation was legal in Texas and Oklahoma until the middle of the twentieth century, and graduate training was restricted to whites only. In 1961, Ervin Perry was the first African American man to receive a graduate degree in engineering at the University of Texas at Austin (predating the racial integration of the football team by almost a decade). That the main pipeline to jobs in the oil industry passes through schools with this racist legacy more or less guarantees a mostly white scientific work force.

Granted, if the oil industry were to recruit at Historically Black Colleges and Universities (HBCUs), they would have limited options. Currently, only six HCBUs in the United States offer an undergraduate geology degree, and only one—North Carolina Central University—offers a master's degree in geology. (In contrast, there are 74 HBCUs that offer degrees in sociology.)[6] Tellingly, the website for the NCCU geology program does not mention training for careers in the oil and gas industry.

A range of professions began requiring advanced degrees in the twentieth century partly because these credentials were obtainable only by whites—a form of institutional racism. In other words, advanced educational credentials once served the purpose of social closure, in which the dominant groups (white men in this case) came to monopolize good jobs for themselves (Weeden 2002).

Barriers exist to African Americans and other underrepresented minority groups throughout the oil industry, and not just in the geosciences and engineering. Federal and state governments subsidize the oil industry in part for its promise to bring "good jobs" to the working-class communities where it extracts and processes oil. In-depth reporting by journalist Alan Neuhauser (2018) writing for *US News and World Report* revealed that African American and Latinx men and women rarely benefit from these promised jobs; their numbers are fewer and their pay is lower than that received by white workers. Moreover, nonwhite people are not welcome in some of the places where oil and gas are pumped out of the ground. In many parts of Texas and Oklahoma, confederate flags are a common sight, constant reminders of the virulent—and hegemonic—racism of the white people who live in these areas. Of course, racism is not limited to certain parts of the country. During a recent oil boom, refineries in Pennsylvania imported workers from Oklahoma and North Dakota instead of hiring from the racially diverse local population (Neuhauser 2018).

Even in Houston, the center of the US oil industry and one of the most racially diverse cities in the country, minority men and women lack equal access to oil industry jobs. Until the 1960s, refineries in Texas excluded women from all jobs, while the men's jobs were segregated by race. Once federal law mandated an end to job discrimination, companies started hiring minority men, but by then, these jobs were stripped of benefits and security, turning them into what sociologists call "bad jobs" (Priest and Botson 2012; Kalleberg 2011).

African Americans have fought an uphill battle for access to top jobs in the oil industry. The American Association of Blacks in Energy (AABE), founded in 1978, is a small group dedicated to promoting the careers of Black professionals in the oil industry. At its founding, this group of a dozen members worked with the NAACP and other civil rights groups to ensure that federal energy policy included their voices. Over their history, these organizations have not always been allies, however. The NAACP, for example, promotes environmental and climate justice, including direct action to shut down toxic facilities, putting it in conflict with the proindustry AABE, which lobbies for the expansion of fossil fuels.[7]

The NAACP opposes the oil industry because of its record of environmental racism. The oil industry is notorious for polluting the neighborhoods of racial and ethnic minority groups. An EPA study found that race—and not poverty—was the best predictor of exposure to oil industry pollution (Mikati et al. 2018; Taylor 2014). Research economists Michael Ash and James Boyce (2018) found that minority communities do not benefit with an increase in jobs when oil facilities locate nearby—in fact, they suffer from increased pollution. In a radio interview, Ash gave the example of an Exxon/Mobil refinery in Baton Rouge, Louisiana, a particularly toxic facility in an area of the country known as "Cancer Alley." Although mostly African American neighborhoods surround the plant, the parking lot at the end of the day is "all white guys getting in their pickup trucks and driving far away."[8] In other words, according to Ash, the jobs in the refinery are not going to the people who live in the area and suffer the most from its pollution.

Environmental racism could repel African Americans and members of other minority groups from taking jobs in the oil industry. It makes sense that people might be reluctant to consider working for an industry responsible for polluting their neighborhoods. My interviews, however, suggest that companies can be quite persuasive at overcoming resistance to their offers of employment. Many of the white scientists I interviewed said that they were aggressively recruited despite their negative impressions of the oil industry. They thought they would never work for the "evil oil industry"—until they were offered lucrative scholarships and internships. Presumably, the right financial incentives could also motivate students of color to overcome any potential resistance to working for oil.

For example, one white geologist who originally "had absolutely no intention of ever working for an oil company" said were it not for financial pressures, he never would have been tempted:

> Unlike a lot of other grad students, I was married and had a child in grad school. And I had a friend who was in a similar boat, and he said, apply for every scholarship you can. And so when the department would send out scholarship notifications, regardless of whether I qualified for them, I applied. So GOG was offering a scholarship and I applied for it and got accepted. And they had a bunch of enrichment events and other things that were associated with the scholarship. Then they offered me an internship. I came and did an internship. The internship turned into employment and I'm working in the oil industry now (laughs).

Oil companies are willing to spend lavishly to convince white students to join their ranks. They could expend similar resources to convince students of color to join their ranks. Instead, oil companies target their recruitment at predominately-white institutions.

The reluctance to reach out to underrepresented racial/ethnic minorities may have to do with political assumptions that recruiters make, although this is speculative on my part. It may be the case that oil companies perceive mainstream environmentalists—who are mostly white—as more easily coopted than those associated with the environmental justice movement, which has a more diverse following. I suspect that companies may perceive it easier to convince white people that the oil industry has limited culpability for the damage it creates than it is to convince racial/ethnic minority groups of its benign effects. For example, companies can frame the despoilment of oceans as the result of "accidents"—an argument that is less persuasive when explaining why companies intentionally site toxic facilities in predominately Black neighborhoods (Bullard 2019; Taylor 2014).

But I can only guess about this, since I only talked to white environmentalists who were hired by GOG. Only two Latinx geoscientists and not a single Black geoscientist were hired by GOG in the five-year window of this study, and none volunteered to participate in the interviews.

In addition to luring my respondents with high paychecks, the oil companies did a good job convincing these future professionals that they valued scientific research. The oil industry scholarships they received as graduate students came with no strings attached, at least from their perspectives. This was also key to overcoming resistance to industry jobs. Thus, another geoscientist explained, his interest in the industry developed thanks to an annual industry-sponsored research symposium at his graduate school. Oil company representatives attended this event and gave him valued feedback on his research. In another example, a major oil company came through with funding for a student's doctoral research:

> My advisor didn't have funding for the particular research I was doing. So through channels—because the university where I was

getting the doctorate had petroleum recruitment—the chair was able to get me an internship that summer. And it paid well. It was interesting because it was something I hadn't expected. They were actually doing geological research. Granted it was toward petroleum exploration, but they were doing research. It was an interesting project, I had a lot of fun doing it, and it kind of opened my eyes to "Hey you can do science and research—maybe not so esoteric—in the industry."

These geoscientists told me that to consider jobs in the industry, they had to be convinced that the oil companies valued scientific research, albeit not necessarily on the "esoteric" topics they had pursued while in graduate school.

Internships seemed designed with this purpose in mind: to convince graduate students that the company valued and respected their primary identity as scientists, as evident in this example:

So I was in grad school for paleo. And the first recruiters that came by, I was like, I am never going to work for an oil company, but I went to the interview for practice. It was only a 15-minute interview. In the end, they must have seen something or liked what I said because they offered me an internship. I went, Well, the money is great (laughs). And, you know, it's three months, I'll give it a try. I thought that when you work for a big business like that you don't get to do real geology; you don't get to test hypotheses. I wanted to do more fieldwork, not be stuck in an office all day. But I did the first internship and loved it. I had so much fun. I got to test a bunch of hypotheses. There is a business aspect to it. But it's/ I want to say, it's almost minor. It's not minor/ any job has things you are not going to like about it. And so that was my first introduction to that.

Although this geologist does not like the "business aspect" of her job, she considers it a minor part of what she does. Minor enough, that is, to assuage her concerns as a scientist working for the industry. She enjoys testing hypotheses, which she did as an intern, and still does today. From her perspective, the industry values science, which is why she entered and why she stays.

Similarly, another woman geologist told me that she was originally skeptical about the oil industry, but ultimately was convinced that there was no better place to be a scientist. The cutting-edge technology along with the vast amount of data available to her as an oil industry scientist overshadowed the paltry resources available to university researchers. However, the biggest drawback of working for the industry is that results are expected instantly, with little time for analysis:

> I tell people, if you are a perfectionist go to work for the oil industry and they will beat it out of you.
>
> [CW laughs.]
>
> I'm serious. You get a week to analyze more data than you've ever seen in your life. In grad school you're beating like ten points of data to death—"I swear I can get four papers out of this." Here, you're like, oh my God, I'm so wrong about everything I ever did.

The trade-off, as she perceived it, was perfect analysis of imperfect data (in academia), or imperfect analysis of excellent data (in industry). As a scientist, she preferred the latter.

Several geoscientists underscored to me the importance of maintaining their primary identity as scientists despite working for the industry. They recalled thinking in college that science and industry were in conflict with each other. These concerns had to be mitigated in order to convince them to work for the industry.

In the following example, a geoscientist told me that she works for GOG precisely because the company enabled her to conduct research. As a graduate student, she had no interest in working for the industry, except, she remembered,

It's sort of a joke I would make because, I said, I have too many student loans to pay off, I can always work for the oil industry. It was kind of a joke, I would just joke around with my friends. Because when you are in academia, especially in my field [oil spill abatement], actually the mentality is if you go to work in industry, you've sold out.

Even though she attended an oily school, a number of her professors and fellow students shared her perception that the oil industry did not respect science; working for them was tantamount to "selling out" one's scientific values. However, near the end of her doctoral program, she got the opportunity to investigate an oil spill. Oil companies controlled the site and previously had denied access to her and her professor despite their National Science Foundation funding. She suggested, "I think that's because of the lawyers. They were probably getting sued. I don't know the whole story there and I don't want to misspeak on that." However, an alumnus from her PhD program who worked for GOG found out about her research. This man approached her advisor with an offer to sponsor her work on the spill once she graduated. This research opportunity eventually morphed into her regular job. In her case then, the opportunity to continue her research overcame her initial reluctance to work for the industry. It resolved the conflict she perceived between industry and science.

Her story also highlights the tight university networks that characterize the industry. At AAPG conferences and other industry events, it is common to see seasoned professionals wearing class rings and alumni insignia on their lapels, displaying the unmistakable dominance of Texas and Oklahoma schools among geoscientists in the oil industry. Thanks to the alumni network at her university, with its close connection to industry, an otherwise inaccessible research opportunity opened up for this respondent.

By the time of our interviews, those who had originally rejected the option of working for the industry had made peace with their decisions. Only one geologist expressed continued ambivalence about her career. She explained that she grew up on the east coast thinking the oil industry was evil, a widely shared view among her close network of friends and family. Her best friend from childhood refuses to speak to her now because she works for an oil company. Her story explains how GOG managed to overcome her resistance.

While in graduate school, she explained, she "rejected a couple of offers for internships." "I would interview," she said, but would always tell the companies "I'm not really interested." Then a recruiter approached her from GOG and expressed interest in her research on an Indonesian earthquake: "He wanted to read my report, and he actually read it and we had some really good conversations about it." They corresponded over the course of two years while she was in school, a relationship that culminated in an offer of an internship in Alaska. "Well," she thought, "I'd love to go to Alaska for the summer."

While in Alaska, she worked with a mentor who "held my hand rather tightly all summer because I had no idea what I was doing. I had no background in oil and gas." She appreciated his support but still had unresolved issues about working for an oil company:

> One day I pulled out my self-righteous Boston cap, and I [said to him], "You guys ruin the environment and kill animals." Sitting here, thinking back on that, I'm kind of embarrassed. The next day, he gave me all the three-ringed binders of regulations for acquiring seismic data and drilling wells in Alaska. It filled up my desk,

basically. And he said, "You don't need to read this but you do need to understand that we are not unethical jerks. We are not doing these things that you've been told that we do. But you prove it to yourself." And so I spent a lot of time proving it to myself. And I feel very comfortable about how this business unit is run. And very comfortable with the industry I guess. Not every aspect of it, obviously.

She now believes that her previous views were ignorant, as are the views of most of the environmentalists she knows, especially those living on the east coast:

> Boston specifically has a history of ignorantly—or non-informed-ly—making assumptions about how evil the oil industry is. Way back when, they condemned slavery, which/ rightly so. But they were still importing molasses, which was using slave labor down in the Caribbean. So like, this is stuff that I have soul-searched about. Because I came to graduate school with a "the oil industry is evil" attitude without being able to defend it. Because I was uneducated.

The respondent draws a complicated analogy between the oil and gas industry and slavery. The point of the story, I believe, is to explain how good people unintentionally harm others if they are uneducated. The Bostonians of old should have stopped importing molasses, since it was produced through the evil institution of slavery, but they were uneducated. In contrast, contemporary Bostonians protest fossil fuels because they think that the oil industry is evil, but they, too, are uneducated. They fail to understand that the professionals who work for GOG are ethical individuals who abide by environmental regulations. Thanks to a devoted mentor who educated her, she now understands that the oil industry is not evil. Although she is not comfortable with everything the industry does, she is confident that her company (or at least her unit within the company) does not pollute the environment. If only the Bostonians of today would understand that too, she would still be in contact with her best friend from childhood.

It is remarkable to me that GOG expended so much energy to convince this committed environmentalist to join their ranks. Why would an oil company do this? I cannot comment on her skills, but I suspect that her gender had a role to play. It may be the case that the company tried hard to convince her because she could help the industry project a cleaner and more ethical image. This interpretation is suggested by the work of sociologist Shannon Elizabeth Bell and her colleagues (2019), who studied the advertising sponsored by oil and gas and other toxic industries. They found that these industries often cast women in the role of company spokesperson when addressing environmental concerns. Gender stereotypes—particularly the assumption that women care more than men do about preserving and protecting the environment—endow women with the cultural ability to purify a tarnished industry. That these stereotypes of femininity apply specifically to white women in society sheds additional light on the racialized recruiting strategies of the industry. By mobilizing white women for these ad campaigns, the authors argue, they essentially protect the companies' power to pollute.

Oil companies care deeply about their public image. A geologist I met at a conference told me that Exxon/Mobil supplied her with a wallet-sized card telling her what to say if confronted by an industry detractor—this after the catastrophic Exxon Valdez oil spill in Alaska. (She said she found the card helpful.) This may help to explain why GOG worked so hard to recruit my respondent. As a white woman, she is a valued spokesperson for the industry's efforts to brand itself as an appropriate steward for the environment—a role she ably performed over the course of my interviews with her.

So far, I have been focusing on the narratives of geoscientists who had to overcome their initial reservations about the oil industry prior to taking a job with GOG. Not everyone I talked to needed to be convinced. For those who had family members in the industry, it was not a particularly hard sell, although they expressed concerns about the industry's periodic boom-and-bust cycles, a topic I will return to later in the book. The influence of family members is yet another gendered and racialized pathway into the industry, since fewer underrepresented minorities compared to white people have a filial connection in the oil industry.

Others wanted to work for the industry for the chance to travel. One male geologist told me that the "opportunity to travel the world and live in different places got me interested in oil and gas." A high school trip to the Middle East exposed him to a desirable ex-pat lifestyle that he was eager to pursue. His positive experience there almost certainly reflects a particular gendered vantage point, but both men and women expressed to me this interest in travel.

The engineers in my sample also articulated these motivations to work for the industry, including family influences and the desire for international travel. However, unlike the geoscientists, the engineers that I interviewed did not express concerns that the industry is evil. I suspect that this is in part because their education prepared them for careers in industry. In contrast to the geoscientists, engineering professionals learn to describe their career choices in ways that contribute to capitalist enterprises (my thesaurus lists "engineering" and "industry" as synonyms). Their professional ethics promote the view that serving the public is compatible with generating profits for their employers. Unlike the scientists I interviewed, the engineers rarely expressed hesitations about working for a toxic industry, reflecting a difference in the ideal worker norms in these two disciplines.

CONCLUSION

Getting more women into science and engineering fields is a national priority. Tapping women's talents is an obvious solution to the shortage of STEM workers in the United States. Although labor economists debate the notion that job openings in STEM outnumber qualified workers, few policymakers and educators question the wisdom of increasing the number of women graduates in these fields. Women are so outnumbered by men in most science and engineering disciplines, they are the logical place to look to fill in any perceived gaps. Many feminists believe that encouraging more women to enter these areas will have the added benefit of diminishing sexist stereotypes about women's capabilities.

The geosciences are one of the few STEM disciplines in which women have made major strides. Women now make up over 40 percent of graduates at all levels. What advice can other disciplines take from their stories of how they entered and succeeded in these fields?

The women geoscientists I interviewed for this study said that they chose their major because they love nature. This is the dominant narrative in the discipline, shared by men and women alike. It appealed to their nostalgic images of childhoods spent in the great outdoors. This same narrative, however, may be a barrier to African American, Latinx, and Native American men and women. The brand of environmentalism that draws many white students into the geoscience major may alienate students from these backgrounds, whose history and experiences of the outdoors conflict with this mainstream narrative.

Entering the geosciences was an easy decision for those I interviewed, but entering the oil industry was not. Having a primary identity as an environmentalist and a scientist is incompatible with working for "the evil oil industry," I was told. However, generous scholarships and internships funded by the oil companies were able to overcome their initial hesitations. Not surprisingly, massive debts from college loans can make these offers irresistible.

These lucrative opportunities to intern and eventually work for the oil industry were available to those who attended "oily schools" funded by the oil and gas industry. These predominately white institutions are the places where companies distribute scholarship money and recruit interns. That the companies target their recruitment at these schools guarantees them a predominately white work force. Although the industry is white male dominated, white women are valued because of the role they can play in symbolically purifying the industry.

Of course, the accounts I gathered are all success stories. I only talked to people who graduated from college and secured job offers at one major oil and gas company. I did not hear from anyone excluded from the oil and gas pipeline, nor did I hear from anyone who washed out in college or who rejected offers of oil industry employment. However, my respondents do shine light on the pathways of white men and women into this oil company. What happens to them after starting their jobs is the topic of the next chapters.

CHAPTER THREE

The Stayers

If getting women into STEM fields is a challenge for many disciplines, as the previous chapter discussed, keeping them there is also a challenge. Once employed, women's attrition from science and engineering careers is very high. Women in STEM occupations are significantly more likely to leave these fields compared to men, especially in the first five years of their careers. They also leave in much higher proportions than women leave other professions (Glass et al. 2013; Harris 2019; Hunt 2016).[1]

Policy makers care about women's attrition from STEM careers because it undermines efforts to promote gender diversity in science and engineering. Companies care about attrition because it represents an economic loss for them. Major oil and gas companies invest a great deal in the training of newly graduated scientists and engineers. At GOG, new hires are rotated through the major divisions of the business. Geoscientists cycle through six- to nine-month appointments in the areas of development, exploration, and operations. The goal of this two- to

three-year training period is to make them well-rounded
employees, and to help them to identify the particular niche in
the firm where they can make their greatest contribution. When
scientists leave after acquiring this training, the company loses
this considerable investment in them. To companies, it does not
matter if their employees are leaving STEM fields altogether or
if they are going to work for a competitor. They simply want to
stop the attrition from the company.

This is why my colleague Chandra Muller and I designed
this study: to identify the factors that led newly hired scientists
and engineers to stay or leave GOG. With the help of the company, we contacted every newly hired STEM worker—over 360
individuals—and asked them to complete a survey about their
career aspirations, including the likelihood of them staying or
leaving their jobs. The goal was to identify what was occurring
in those critical first five years of their careers when STEM
workers are most vulnerable to leaving their jobs.

We did not detect much attrition during the first three years
we followed them, but by the fourth year, scientists and engineers started leaving in large numbers. This is when oil prices
began to fall. By the fifth year, 40 percent of the original sample
had left GOG, including 50 percent of the women, and 36 percent of the men. An astonishing 75 percent of women with young
children left GOG by the end of our study.

I conducted in-depth interviews with volunteers from the
larger study, including 22 men and 18 women whom I followed
over a three-year period. I witnessed even more extreme patterns of attrition in this smaller sample. Half of the men I followed left the company by the end of the five-year period, while
72 percent of the women left, including all of the mothers of
young children. (See the methodological appendix.) Among the

14 women geoscientists I followed over the years, only three remained working at GOG.

In subsequent chapters, I will explore the narratives of those who left the company. Here, I focus on the accounts of the three women geoscientists who stayed. I will refer to these women as "successful" because they are among the select few that the company retained during the period of downsizing. I recognize, however, that success comes in many forms, and that those who left may be successful in ways that have nothing to do with GOG or their scientific careers. I am interested in understanding what these three women share in common that might explain their staying power at the company. Although I cannot comment directly on their skills, I can learn from their narratives what the company values in its employees. Listening carefully to their accounts of their experiences over time, I uncover features of their workplaces that enabled them to stay while many others were leaving.

Focusing on the narratives of successful scientists carries a number of interpretive risks, however. First, it is impossible to explain the determinants of success by exclusively looking at successful cases, a logical fallacy that researchers call "selecting on the dependent variable." Although successful women scientists can offer helpful insights about coping and even thriving in male-dominated environments, they cannot identify why they succeeded while others failed.[2] Moreover, it can be tempting for them to explain their success in self-serving terms. Successful people often emphasize their outstanding personal qualities, including their merit, tenacity, and passion, as the keys to their accomplishments. While all of these can be critical for success, it is important to keep in mind that many people who leave these careers might also be meritorious, tenacious, and passionate. We cannot fully understand what it takes to succeed without also examining the experiences of those who leave (which is the topic of subsequent chapters).

A second risk of using career narratives to explain patterns of retention is that workers are not always privy to the reasons why they succeed or fail at work. An employee's fate is in the hands of managers who decide whom to hire, promote, fire, or lay off—decisions that typically happen behind closed doors. Officially, GOG told the employees that success depends on two factors: *merit* and *skill set*. *Merit* is assessed in yearly performance reviews, when each employee is rated by their supervisor on a scale of one to five, with five indicating a "top performer." Those at the bottom of the scale lose their jobs. *Skill set* refers to one's expertise and experience in the organization. According to GOG, the top performers whose areas of expertise perfectly align with the company's needs are the most successful employees. In the narratives that follow, respondents weigh these official pronouncements against their personal experiences at work as well as the experiences of their coworkers. Their accounts sometimes confirm and sometimes challenge the official corporate line. In the end, however, they can only guess why they are the "chosen ones"—an important consideration when interpreting their narratives.

There is a third, more perilous risk of focusing on the stories of successful employees. Companies may use this information to decide whom to hire in the future. This was a concern of mine in this study: By ascertaining who was staying and who was leaving, I feared that GOG might alter its recruiting practices and hire only those people who matched the profiles of its most successful scientists. In the worst-case scenario, if mostly white men were successful, this might encourage the company to hire only white men! Obviously, I wish to avoid this outcome.[3]

Instead of reinforcing such preferences, my goal is to investigate how the company fosters different career outcomes for different groups. In other words, I examine the career narratives of these successful employees with an eye to understanding how work organizations identify and reward "top performers." Their accounts help me to understand what the company values in its scientific workforce. Talking to the stayers, I am able to identify the hegemonic beliefs about what it takes to thrive—or at least survive—in this corporate environment. As I will show, the narratives of the stayers also reveal the individual strategies that scientists use to navigate obstacles to their careers.

In what follows, I examine the career narratives of the three women geoscientists in my study who remained at GOG in order to reveal how advantages operate there. The three women are white, in their mid-thirties, and married (in two cases to men geoscientists also employed at GOG), and none of them has children. All received perfect scores in their performance evaluations. They are highly paid scientists, earning over $150,000 per year. By focusing on their narratives, my goal is to identify what GOG values in its scientific workforce.

LYNNE
"When Cuts Are This Deep, Any Little Thing Makes the Difference"

When I first talked to Lynne, I predicted she was a prime candidate for attrition. On her survey, she gave herself only a 25-percent chance of staying at GOG, and a 50-percent chance of leaving the industry altogether. When asked why, she said that she and her husband (an engineering consultant) hated Houston, a common sentiment among many of the geoscientists I interviewed. Houston is the unofficial capital of the US oil industry, but it is the last place on Earth they ever wanted to live. As self-described "small-town people" who love the outdoors, Lynne was willing to change careers in order to relocate.

Three years later, Lynne was arguably the most successful of the 14 women geoscientists I followed over time. At our last interview, she was working in a challenging new leadership position that she enjoyed, and she was receiving company-wide recognition for her mentoring and recruiting efforts.

Lynne entered the industry almost by accident. She went to graduate school planning to study paleontology, but became more interested in rocks than dinosaurs, she told me. Never wanting to work for the oil industry, she attended a recruiting session on a lark and received an offer of an internship after a brief interview. Two more internships followed, all at major oil companies. She ultimately chose GOG because it had offices outside of Houston; she hoped she could transfer to one of their satellite locations after a few years.

A secondary concern that tipped the scales in favor of GOG was the company's approach to career development. Major oil companies, like many large firms today, have adopted the model of "career maps" for their professional workforce. Career maps, sometimes called "I-deals," are individualized programs of career development. In many firms, they have replaced career "ladders," which set out standardized and uniform pathways to promotion through the company. A career map, in contrast, establishes individualized goals and expectations that managers use to monitor workers' productivity and evaluate their performance (Williams, Muller, and Kilanski 2012).

Career maps also guide internal transfers. At GOG, geoscientists change areas frequently, every six to nine months for the first three years and once every three to nine years after that. Employees decide on their future placement in consultation with their supervisor and with their assigned "talent manager" at the firm. Lynne was attracted to GOG, she told me, because the company would assign a geoscientist as her talent manager, a far superior arrangement than those offered by the other majors, which typically assigned engineers to this role.

Lynne's account of her earliest placements at the firm illustrate how career maps are designed to function:

> When you first come here, a role is assigned to you. And that comes down to the hiring committee and what positions supervisors have told them are available. They try and match all the incoming people in that year to what positions might be available.... So for me coming in, I said in my interview, "I'm very detail oriented; I like to work with a lot of data." That is our description for a development geologist (laughs). So because of the things I said in the interview, they put me in a development geo[4] position. But our next rotations are decided by what we call the talent management team. And that's you communicating to them what you want to do next. Whether you want to become a specialist, whether you want to be a generalist. Whether you like exploration, development, operations. You work with your supervisor and your TM [talent manager] to find your next position.

Lynne describes a number of important features of career maps. First, career maps are tailored to the individual's unique talents. In Lynne's case, it was her expressed proclivity for detail work that resulted in her first assignment. Second, career maps give employees responsibility for managing their career development. Lynne emphasizes that it is up to each person to identify their niche within the firm by seeking out internal transfer opportunities. Third, career maps depend on employees effectively communicating their aspirations to their supervisors and talent managers. All three parties must work together to find the right placement.

In some respects, this individualized approach to career development is an improvement over the "career ladders" of the past. At GOG, it was often said that employees "drive their own careers"—an arrangement that many preferred to the lock-step, one-size-fits-all alternative. However, organizing a large bureaucracy to cater to and cultivate everyone's particular interests and talents presents a mammoth organizational challenge. It seems practically impossible to implement this business model for 20,000 employees. Lynne acknowledged that the process does not "work for everyone," but it had for her because she has been proactive in developing an extensive network to identify future opportunities. She said she also cultivated positive relationships with her supervisor, "who recognized something [in me] and gave me an opportunity, and I've excelled at it." In fact, at the time of our first interview, Lynne was already a team leader, the first level of management in the company. From her perspective, those who excel at GOG are able to cultivate positive relationships with superiors, communicate effectively, and develop a strong network throughout the company.

By the time of our second interview, company layoffs had begun, and Lynne had lost her leadership position (along with her supervisor, who had retired) and had been transferred to a new specialty that she did not enjoy as much as her previous one. The company eliminated her entire team during early rounds of downsizing, but managers elected to move Lynne instead of laying her off, which is the fate that befell most of her coworkers. How did she survive? Echoing the official criteria that I mentioned earlier, she told me that her flexibility to go where the company needed her and her willingness to abandon her career map made a difference. Gone was any talk of "driving your own career."

From Lynne's perspective, two additional things helped her survive the layoffs: her optimism and her work ethic. Throughout the downsizings, she maintained a relentlessly cheerful attitude, she said. She joked with her colleagues that losing their jobs would mean they finally could take that dream ski vacation or visit all the National Parks. "We're not going to be homeless," she offered, reassuringly. In her own case, she thought she had a 50–50 chance of being let go, something she felt she could not control because "I either fit in the organization or I don't." It is just not worth worrying about, she told me. However, to prove her worth, she worked extremely long hours, often taking work home to keep up with an ever-increasing workload, even "cranking out paperwork while watching TV." She cut back only when her husband started complaining that she was ignoring him.

From her perspective, having an "above and beyond" work ethic combined with a positive attitude made her a great fit for the company, but she could only hope that her superiors felt likewise. By the fifth year of the study, it was clear to her that they did. After surviving three rounds of layoffs, Lynne was promoted to a supervisory position in yet another specialty area not of her choosing, but one that she said she enjoyed.

At the end of the study, Lynne was optimistic about her future at the company, but she was no Pollyanna. Over the years of interviewing her, she shared concerns with me about the company's lack of gender equality. Lynne knew she was the exception and not the rule, as she was fast becoming one of the most senior women geoscientists at the company (at age 37!). Once the layoffs began, she could not help but notice the steep decline in women's numbers. The company denied that its diversity had changed at all, but Lynne suspected that they were comparing the current percentage of women to what it was 15 years earlier, before any diversity initiatives began. Lynne was certain that women were targeted for layoffs, especially those with young children. To illustrate, she recounted a story told to her by a woman colleague:

> On her floor, the nursing room before the layoffs was constantly busy. "And now," she said, "there's three of us." That is such an impactful statement. "It used to be full. Now there are three of us." And it's not because all of those women stopped nursing. It's the layoffs.

Lynne speculated that mothers were vulnerable because they worked fewer hours on average. In deciding whom to let go, the company favored people like Lynne who were taking on additional responsibilities, not those who were working reduced schedules. She lamented,

> Anyone who was working part time—and it was mostly mothers who were working part time—they were hard hit. Mothers were hard hit.... When cuts are this deep, any little thing makes the difference.

Here is the essence of a gendered organization: workers who devote themselves unconditionally to work are favored employees. Although scholars usually associate this "ideal worker norm" with married men whose wives look after their house and children for them, Lynne has managed to fulfill this expectation. With a supportive spouse and no children at home, there is not a single "little thing" about her that distinguishes her from the corporation's ideal worker.

Actually, there was one thing that distinguished Lynne. She had become an outspoken advocate of the company's new "inclusion" initiative. She explained:

> There is a big push now for, not so much diversity, but "inclusion." Diversity of thought and mind and all sorts of diversity, not just diversity of gender, races.

"Inclusion" has become a popular buzzword in organizations, adding to, and in some cases supplanting, the short-lived efforts to promote diversity. Diversity became a high-profile issue in corporations starting in the 1990s, in response to discrimination lawsuits. Corporations publicized their commitment to gender and racial diversity in their mission statements, job advertisements, recruitment materials, public relations, and personnel policies. A cottage industry of consultants developed to help companies achieve diversity, advertising their services as a means to improve the corporate bottom line and reduce potential legal liabilities. In response, most major corporations instituted a variety of diversity management initiatives; some of the most popular of these include affinity groups, formal mentoring programs, diversity training, and targeted recruitment and promotion programs (Berrey 2015; Williams, Kilanski, and Muller 2014).

On the surface, corporate efforts to promote diversity seem promising. However, after two decades of the corporate "diversity craze," executive suites remain overwhelmingly white male–dominated. Sociological researchers failed to detect much change in the leadership of companies that instituted diversity policies (Kalev and Dobbin 2006). The one program shown to be effective—Affirmative Action–type programs that establish target numbers of women and minority men for recruitment and promotion—were rarely enacted and always controversial, even among several of the white women I interviewed who were their primary beneficiaries (Williams, Kilanski, Muller 2014; see also Hirsh and Tomaskovic-Devey 2020).

Today, diversity programs have merged with or have been replaced by "inclusion" initiatives. So what is the difference? As Lynne explained, GOG's inclusion initiative seeks to recognize and respect all sorts of differences among employees, "not just gender, races." People with different personalities, past experiences, and professional specializations all promote inclusion and enhance the corporate bottom line.

At an industry conference on women's leadership that I attended,[5] executives from major oil and gas companies extolled the value of inclusion, suggesting that it had become the industry's new orthodoxy. Morag Watson, Vice President at British Petroleum, explained to the audience of mostly young women scientists and engineers of different racial/ethnic backgrounds that the focus on diversity was misplaced: "If you start with diversity first, you're dead. Start with innovation and inclusion, you will build a culture that values people—they will come." In her view, inclusion is the necessary precondition for developing a diverse team. The more inclusive your team, she maintained, the more innovative it will be. Only then will your team attract diversity—women in this context—presumably because they will clamber to participate in such innovative work.

This claim is riddled with many dubious assumptions, including the notion that women can choose which teams they work on. The "career maps" that were instituted to allow workers more autonomy in their placement were abandoned during the downturn. This was Lynne's experience: she went where she was told to go.

More perniciously, this focus on inclusion absolves managers from hiring women. Not a single corporate spokeswoman on this six-person panel at a conference on women's leadership advocated hiring more women into the industry. Instead, each woman said companies should pursue inclusion. Shamilah Saidi, an executive at Petronas, stated boldly, "We value diversity in thinking." She then referred to a trip she made to Silicon Valley, which changed her thinking about diversity. I sat in the audience dumbfounded that anyone would consult with Silicon Valley about diversity—the tech firms there are notorious for discriminating based on gender and race. Upon reflection, however, it makes perfect sense that this would be the source of the new emphasis on inclusion. Inclusion shifts attention away from easily counted demographic categories to invisible qualities that elude measurement. An inclusive workplace contains people with a range of ineffable qualities that reach deep inside the individual to the level of neurons and synapses, Saidi explained. With great enthusiasm, she recommended the audience read Scott Page's book *Diversity Bonus* to understand why diversity in thinking matters more than gender and race. She explained that Page, a mathematician, proves that "cognitive diversity" enhances "nonroutine decision-making"—goals that cannot be achieved by increasing representation of gender and race in the wrong-headed pursuit of "social justice."

Whether or not these claims about Page are correct, it is remarkable how this approach is reshaping the discussion about social inequality in the white male–dominated oil industry. While the older focus on diversity was not successful at dislodging the white male control of the industry (which was arguably never its intent), it did make companies accountable for their lack of gender and racial representation among executives and board members. This new approach to inclusion sweeps away these metrics. The range of differences included under the rubric of "inclusion" is so broad as to be meaningless. Instead of rectifying the underrepresentation of women and minority men, companies can accomplish the mandate to include the "diversity of view-points" with a room full of white men (Wingfield 2019:32).

Lynne believed in inclusion, though. She thought that by promoting inclusion GOG could express support for those who survived multiple rounds of layoffs. She said,

> I think that [inclusion] helps with team dynamics: "I'm a 30-year-old white man. Who's looking out for me?" Well if you say "inclusion," that includes you as well. Diversity, you might feel excluded, like "Ah, I'm going to get set aside because of that."

With her characteristic energy and optimism, Lynne took it upon herself to establish a "cultural inclusion" program at GOG. She now works (on her own time) to convince her colleagues that they have a stake in promoting corporate inclusion. Her program sponsored one event to highlight the differences between introverts and extroverts, promoting the case that both personality types deserve a place at the table. Her program also provides training in unconscious biases, with the goal of making everyone more accepting of others. She explained:

> Anything can be taught. Most of us are scientists. We know how to learn. But we don't know what we don't know.... A lot of it is unconscious biases. One thing that I learned, I had no idea about/ women tend to judge other women harsher, in terms of the work that they put out. OK, that is a bias that I may or may not have. I have to watch out for that. Am I judging this person harder? And all you have to do is ask yourself that question, knowing that might be there. Learning a lot more about your unconscious biases, then you can work, you can make this better. I think any professional development interpersonal skills can be taught.

The inclusion initiative at GOG has flipped the script on how companies address gender and racial bias. Discrimination is no longer the result of actions taken by managers; it is caused by the unconscious attitudes of coworkers. Although aware that the company targeted young mothers for layoffs, Lynne interrogates her own unconscious bias against women. By arguing that "inclusion" is a teachable "professional development interpersonal skill" that anyone can learn, she identifies personal transformation as the key to women's retention in the industry.

Lynne enthusiastically supports the official inclusion initiative, hoping that educating her coworkers about their biases and emphasizing the value of "diverse perspectives" will promote a corporate climate that will build more support for hiring and retaining women. Paradoxically, however, the climate is not the reason she stayed. Lynne is a survivor because she is committed to the company's success. She is willing to play the role of corporate cheerleader and go "above and beyond" to support corporate initiatives. This is the devil's bargain made by many successful women: advancing the cause of gender equality requires her to take on extra work for a corporate initiative that seems designed to make men, and not women, feel more accepted by her organization.

Sociologist Adia Harvey Wingfield (2019) discovered a similar dynamic among the successful Black professionals she interviewed for a study on racial inequality in the US health care system. She found that these workers assumed responsibility for addressing the underrepresentation of Black professionals at their hospitals, as well as improving health care to the predominantly Black communities they served. She calls this dynamic "racial outsourcing":

> Racial outsourcing occurs when organizations fail to do the work of transforming their culture, norms, and workforces to reach communities of color and instead rely on black professionals for this labor. As

a result ... black professionals do *equity work,* which I define as the various forms of labor associated with making organizations more accessible to minority communities.... In some cases, equity work is explicitly mandated by organizations; in others, it is a result of the racial solidarity many black professionals bring to their work. (34–35)

With racial outsourcing, organizations can abdicate responsibility for creating diverse institutions, and offload this work onto already overburdened Black professionals, who receive no compensation or recognition for this additional labor.

Like Wingfield's Black health care professionals, Lynne is devoted to improving the diversity of her workplace. She participated in my study because she hoped it would help to bring about greater gender equality at GOG. Based on her experience, retention depends on a strong work ethic, an optimistic attitude, and an extensive network—all attributes that she encouraged other women to develop. She pointedly asked me what more she could do to support women. The problem is that Lynne is not in a position to help other women; only her managers are. However, instead of criticizing managers for firing women, Lynne champions what appear to be their cynical attempts to shift the burden onto workers for creating an inclusive work environment. In a stunning sleight of hands, her managers escape responsibility for firing women. Instead, they have outsourced the work of promoting equality onto one of the few remaining women geoscientists at the company.

PATRICIA
"Too Many People and Not Enough Rigs"

Patricia came to GOG after working five years for a different major oil company. There she developed expertise in hydraulic fracturing, or fracking. At that previous company, Patricia was responsible for managing four rigs that operated around the clock. Exhausted by the pace of the work and on the verge of burnout, she switched to GOG, where her geoscientist husband also works. Instead of four rigs, GOG assigned her to a job overseeing only one rig, which from her perspective was a very sweet deal.

Patricia's husband figures centrally throughout her career narrative. Patricia received her undergraduate degree in mining engineering, and then followed her geologist husband to graduate school in another state. Because the university where her husband enrolled lacked a mining engineering program, Patricia shifted to geology for her own graduate training. The graduate school they attended was not defined as an "oily school," so neither enjoyed the benefit of an oil industry internship. However, GOG offered her husband a job after he graduated, and Patricia followed him to Houston. Patricia found a job there at a small independent producer but a promised position never materialized (a case of "bait and switch"). She found her next job through the AAPG student expo in Houston. She qualified as a "student" because it had been less than a year since she graduated. Patricia told me that this was a lucky coincidence because otherwise it would have been impossible for her to get an interview at a major oil and gas company.

Looking back, Patricia describes her career development as "faster" than average. After transferring to GOG, she assumed her first leadership position in three months' time, when she was named "team lead." Her team, comprising 20 scientists and engineers, was drilling an unconventional well in a recently developed oil field (referred to as a "play"). Her team was very young, she told me, with most people in their twenties. "It's kind of interesting," she told me. "If the high-level managers within

GOG ever saw who was drilling their hydraulic wells, they would have some queasy stomachs about it because it's primarily people with five years or less experience." In addition to being very young, her group was very male. She joked that it was a "young boys club" when she arrived. She was the only woman on the team.

Over the course of two years, however, the gender composition of her team changed to 50–50 men and women. "How did that happen?" I asked. After assuming the team lead role, she told me, she made it a priority to bring in more women. "It seems like our team just functions better when there is an even mix of men and women," she explained. At her previous company, there were many more women scientists and supervisors compared to GOG. She remembered walking through the halls when she first arrived at GOG and thinking, "Hmm, this is different." She was determined to change things, and she relied on the network she developed at that other company to do so:

> When an opening comes up, I usually try to think of women that I know from previous experience that may be good for that position. And not only just for our team, but for the other teams that we work with. Any time we're looking for experienced hires, I try to reach out to friends of mine or former colleagues of mine to say, hey this position is open. Are you interested or do you know anybody else?

Having a sympathetic boss also made a difference in achieving her goal of gender balance on her team, she told me. "He's concerned about it as well. It helps when it is not just my voice." Working together, they recruited eight women from Patricia's former employer.

During that first interview, Patricia was more than satisfied with her career, but she only gave herself a 50-percent chance of remaining at GOG in the near future. Her thoughts of leaving revolved around family planning. She wanted to have children, she told me, but she was uncertain how family life would meld with her busy work life. Making matters worse, at GOG, arrangements for maternity leave were entirely at the discretion of supervisors. Although she was sure that her current supervisor would work with her to make suitable arrangements for family leave, she feared what might happen once she changed work groups. The company requirement to move to a new group every few years impeded family planning.

In this regard, Patricia's situation is typical among professional women in the United States. In most states in this country, employers decide whether to provide access to family accommodations (four states are now among the exceptions to this rule), and workers decide whether to take advantage of them. In contrast, the European Union, for example, mandates parents' rights to flexible schedules, part-time schedules, paid leave, and public provision of child care; in some countries (e.g., Iceland and Sweden), policy take-up is mandatory (Collins 2019; Grabham 2014; Herman, Lewis, and Humbert 2013; Lewis and Campbell 2008). Although mandatory policies are not without controversy—some feminists argue that they can undermine gender equality in the labor force—most scholars agree that public policy in the United States is more hostile to working mothers than that of any other country in the developed world (Collins 2019; Glass 2009; Gornick and Meyers 2003; Pettit and Hook 2009; Williams 2000).

Like most US-based oil and gas companies, GOG offers only the federally mandated 12 weeks of maternity leave (six weeks paid disability leave and six weeks unpaid leave), no special accommodations for new fathers, and no childcare assistance.

Also in line with most US corporations (Correll et al. 2014), GOG gives individual supervisors discretion to negotiate flexible schedules with employees. Patricia was among the many women I interviewed at the company who lamented the paucity of these accommodations. She said:

> The thing that I don't like about GOG is so much of the entire maternity flexible work policy is at your manager's discretion. Right now the group that I'm in, the management is fantastic about that. There are lots of options, they're very supportive of whatever your plan wants to be, they'll help work with you. And so if I'm within the same group when we have children, I have no doubt that we'll find something that works. However, that's not always the case. I have friends in other groups, who their boss doesn't get it and says, "Nope, you're here or you're not. I don't want three quarters of an employee." It's hard when there is no real company policy.

Patricia told me horror stories of women promised accommodations that never materialized. In one case, a mother returning to work found out that the company had depleted her vacation time to pay for her extended maternity leave. Patricia was outraged:

> We had one lady who ended up quitting GOG because she came back from maternity leave and they had taken up all of her vacation. She was like, "Nobody told me I would lose my vacation! I came back in April and now you are telling me that I have no vacation until the end of the year?" "Oh well, they told you wrong! Your vacation is gone." So she said, "Fine, then I'll go to another company that will give me vacation!" It irks people enough to say, "Hey if you are not even willing to work with me, I'm out of here." Especially right now in Houston. I mean, the job market is so good that you really don't have to put up with anything that you don't want to.

Patricia contemplated the possibility of leaving GOG once she was ready to have children. Importantly, she did not contemplate opting out of her profession, but rather finding a more accommodating employer. This discredits the stereotype that women willingly sacrifice professional careers "for the family"— a common belief that has been challenged by a number of studies (Damaske 2011; Glass 2009; Stone 2007). Patricia expressed confidence that she could negotiate with her employer for suitable family arrangements when the time came. At that first interview, the price of oil was high and her skills were in demand; she heard from recruiters at least once a week. She had plenty of reasons for optimism.

The following year, the price of oil fell 50 percent and the first round of layoffs occurred. Patricia not only survived the cuts, she moved to a new, sought-after position. She was now in charge of developing corporate "best practices" for drilling unconventional wells (i.e., fracking), a position that drew on the technical and managerial experience she had accumulated over ten years at the two companies. The position gave her enormous exposure at GOG. At age 33, she had become the company's top geologic expert on hydraulic fracturing.

Her technical expertise saved her, she speculated. Her age also helped, she told me, as the first round of cuts targeted older workers. At GOG, everyone over age 50 had the option of "retiring" early, which most people in this age group opted to do, she said. She said she understood the wisdom of reducing the workforce in this "relatively painless" way, yet she worried about its unintended consequences. "I'm not sure that management understands how much we still relied on those senior people," she lamented. Although she was confident in her skills, she wanted the opportunity to consult occasionally with more experienced workers.

Most of the people I interviewed agreed with Patricia, perceiving that workers over age 50 were choosing to retire—a decision that they thought was good for them and good for the company. Granted, my respondents had an interest in the company's decision to target older workers for layoffs, since most of the people I talked to were in their thirties, an artifact of our study design. Only a handful of my respondents challenged the view that retirements were voluntary, including one 55-year-old who indignantly complained to me about being "force-retired."

In the United States, forced retirement is a form of illegal employment discrimination. Age may not be an official reason for dismissal from work (although there are some exceptions, as in the case of airline pilots). GOG thus had a strong incentive to define these retirements as voluntary. As if admitting they were on legal thin ice, the company required their "retirees" to sign a document promising not to sue the company for discrimination as a condition for receiving their severance pay (they required the same promise from everyone who was laid off).

Layoffs often target older workers based on the assumption that they are more expensive to employ than younger workers are. Some scholars dispute this as an ageist stereotype, pointing out that older experienced workers typically compensate for their higher salaries with higher productivity, loyalty, and skill levels (Roscigno et al. 2007). However, at GOG, the salary, benefits, and bonus structure definitely favored workers with accumulated experience at the company. Few doubted that letting go older workers was saving money for the company. The one older engineer in my sample who was not forced out had joined the company after a 20-year hiatus. I asked her to speculate why she was retained after so many others in her age category had been let go. "Because I'm cheap" was her pithy response.

Retirements accounted for most but not all of the staff reductions, according to Patricia. Another consideration was company need, which put young people at risk of layoffs as well. In fact, she told me, the company laid off a third of the members of her previous group because drilling activity had slowed during the downturn. She explained,

> The problem with that group is that they were staffed for an 18-rig program and they were running six rigs. So it was a mismatch. We worked so hard to build the numbers on that team because we were so short-handed for so long. We finally got caught up, oil prices went down, rig activity slowed, and then we went "Oops." And now we have too many people and not enough rigs.

Unfortunately, she said, "all of them were people we had recruited." In other words, GOG laid off the women she had brought in from her other company. "[My former boss] and I were the ones that recruited a lot of those people," she reminded me. "And we thought, 'Oh, man, there goes six months' work trying to find these people.'"

Patricia did not express any resentment to me about how the layoffs unfolded. She understood the necessity of layoffs from the company's perspective.

> I don't think anybody saw it coming. I think if anything, we were really lacking in a lot of unconventional experience. So we had to hire people. I think we were a bit overzealous because we had money floating around, so it didn't take much to justify hiring somebody. And we were not very efficient at the time either. At first it really did take a lot of manpower. But now we're learning that things that used to take three or four people, maybe we can do with one now. Which is good for the company—that's great. But from a those-three-people-are-my-friends-and-now-we-have-to-let-them-go [perspective], it doesn't feel good.

Patricia considered layoffs "good for the company" even though she felt awful seeing her friends lose their jobs. "We have to let them go," she said, implying that this is an inevitable part of the oil business. Her use of the collective "we" also indicates that she identifies with the company's actions, even as she struggles to accept these decisions on a personal level.

Patricia's ability to recast the story of layoffs in positive terms seems key to understanding her resilience at the company. I noticed that she found ways to explain the company's decisions that made them seem reasonable. Thus, the layoffs primarily targeted people who wanted to retire anyway, or the layoffs are an inevitable part of doing business in a volatile industry. Seeing management decisions as rational perhaps made it easier to identify with the company and maintain enthusiasm for her job. Patricia put an additional, even more positive spin on the layoffs to me, speculating that because of the high turnover, more opportunities would open up for women in management. She reasoned that since mostly older people were leaving the company, younger people would begin to assume more leadership positions, and because the younger cohorts contained more women than the older cohorts did, GOG leadership would become more gender balanced once the dust had settled.

I call this *the demographic theory of social change*. It is a common conceit among young people that after the older generation retires, the more enlightened and progressive young people will take over organizations and advance the goals of equality and fairness. Patricia hoped that younger workers would bring increased diversity and an attendant commitment to social equality with them as they replaced the prejudiced older people who retired.

By the fifth year of the study, that theory proved wrong. Successive rounds of layoffs had made the situation with diversity even worse. Patricia lamented:

> That is one of my pet peeve issues at the moment. We lost women in leadership positions. I actually asked HR for the statistics, and they told me that they couldn't get them, and I told them they were lying to me (laughs).

Patricia's laugh expressed exasperation and sarcasm, not happiness.[6] She noted, "I think there is one female manager in the company. There are supervisors that are women. But upper-level managers, I can think of one, which is pretty sad."

Instead of addressing the growing lack of gender diversity, the senior leadership was promoting "diversity of thought," as part of the inclusion initiative that Lynne was championing. Patricia reflected:

> I'm a little bit concerned. We had a town hall and I asked that question [about women in leadership]. [The senior management] said they just had a leadership forum and talked quite a bit about not only diversity in men and women, but diversity of thought. We have a leadership team that is very old, very white, and very male (laughs), and that is probably not representative of the people that work in the company. So I guess there is a big kind of effort that is just kicking off to really try to open things up and start a dialog. I'm hopeful that will go somewhere. Just the fact that at least the leadership team acknowledges it is an issue is a start.

Describing the "diversity of thought" initiative as "a big kind of effort" that will "start a dialog," Patricia once again displayed her penchant for putting a positive spin on matters. When I pointed out to her that this "start" was happening four decades after second-wave feminism put women's leadership on the national agenda, she laughed again, this time at the absurdity of "old white men" touting the importance of "diversity of thought" while firing women in leadership.

In addition to losing women in leadership positions, mothers of young children also left during the downsizing, Patricia told me. Like Lynne, she suspected that their vulnerability to layoffs was the consequence of their lowered productivity while on maternity leave. They simply could not compete on performance reviews with those who were working full-time. Patricia told me that the women who remained were postponing motherhood until they could work out a timing strategy to game the system of performance reviews:

> My women colleagues say, "Well, we were'thinking about having a baby but I really think I need to have a baby in the winter." Honestly, between October and March, nothing happens. Your performance reviews are due in October for the previous year and then you start a new one in March, so it's kind of like the dead time that people forget.

She and her coworkers were hoping to escape the motherhood penalty by giving birth and taking maternity leave in the winter when work slows down, thus avoiding negative performance reviews. However, this involves perfect timing on getting pregnant. "That's not how it's supposed to work," she averred.

Despite her inclinations, GOG was making it difficult for her to recast the layoffs in positive terms. The company's actions stopped making sense, Patricia told me: "By the time we got to that fourth round of layoffs, the people you expected to be gone are gone, and now who do you take?" The company never explained its actions, aside from reciting its mantra that retention depended on merit and skill set, but their decisions were indefensible, she told me. On her current team, two of the five members were laid off, a cut that was especially painful because it was made without the input of their supervisor. She no longer tried to justify GOG's actions to me. She stopped feeling loyalty to the company. "For me now, it's a job. The company pays me, I give them my work." Unlike Lynne, she did not take on the role of corporate cheerleader.

Instead of defending the company, Patricia told me hopeful "after the layoff" stories. These were optimistic accounts of what happened to her colleagues after leaving GOG:

> It has been cool to watch. My husband and I go to the farmer's market. There's a lot of people there that are former oil and gas people. Now they bake, or they make soap. It's just cool to see what people do. Sad as it is when they lose their job, it's cool to see the kinds of stuff that people start doing. It's been kind of nice.

Meanwhile, she and her husband were talking about becoming beekeepers and starting a business making honey for a living; she was attending workshops for women entrepreneurs to explore this possibility. She was now recasting the possibility of her own layoff in positive terms.

Although Patricia continued to thrive in her career, she felt little loyalty to GOG. From her perspective, she survived the layoffs because of her special expertise: she had drilled the most unconventional wells of anyone at the company. However, looking at her narrative in the larger context of this study, she possessed additional qualities that might have saved her from the chopping block. First, her age probably played a role in her retention. Although she earns an enviable salary that puts her in the top 10 percent of all income earners in the United States, she earns less than what seasoned professionals earn—most of whom were "retired." GOG employs mostly young, inexperienced workers in their oil fields because doing so saves the company money—not because they have special technical expertise. Indeed, Patricia feared the consequences of putting 25-year-olds in charge of fracking large swaths of the country because they did not necessarily know what they were doing.

Patricia's family status also seems critical to understanding her retention. Mothers were targets for layoffs, but she is not a mother, at least not yet. Patricia still wants to have children, which she now understands is a career-ender, unless she can time her pregnancies just right. There is also the matter of her husband. Patricia told me that her husband is a "top performer" at the company and that GOG is willing to do anything to retain him. To be sure, Patricia is no slouch, but the corporation's commitment to retaining him could contribute to their decision to retain her.

Spousal hiring and retention programs are uncommon in the oil and gas industry. (In contrast, spousal hiring has become commonplace in academia.) Based on my conversations with several executives in the industry, I have the impression that spousal accommodations are available only to those, like Patricia's husband, who are labeled top performers. A form of corporate favoritism for men, this could help to explain why women scientists married to other scientists are the women least likely to leave careers in science (Glass et al. 2013). At GOG, they may have been protected from layoffs in an effort to retain their "high-performing" spouses. Notably, this preference for hiring the spouses of successful men also is a form of racial privileging, since few minority men occupy top scientific posts (Williams 2021).

Patricia's survival at GOG is the outcome of employment policies that discriminate based on age, motherhood, marital status, and race. She struck me as a kind and optimistic person—in many ways an ideal employee. However, her narrative reveals the workings of a gendered organization that has enabled her to survive at the expense of many others who left.

EVE
"I Became a Grouchy Sad Person"

Family relationships also figure prominently in the narrative of the third and final "stayer" in my sample, a woman I call Eve. Eve met her future husband at GOG. After they got together, his career played a major role in her career trajectory. When I first talked to Eve, she was in Australia drilling an offshore well. At age 30, this was very early in her career to score a coveted overseas assignment. Many of the geologists I interviewed longed for assignments outside of Houston, let alone outside of the country, but most had to wait several years before becoming eligible for these transfers. When asked how she got this position, Eve explained:

> My partner got offered a job here. He said he would take the job if they found something for me as well, and they did. We both had a desire to get out of Houston for a bit. With him getting offered this position, it was definitely the right thing to do. But for me, yeah, I was a bit of the tag-along in that situation (laughs).

For Eve, this placement was a mixed blessing. She loved the expat lifestyle, especially the travel opportunities that came with living abroad. Unfortunately, she was not at all satisfied with her job.

Eve raised many concerns during our first interview, including feeling a lack of coordination on her team and difficulty communicating with her supervisors. She said that they tell her to assume more initiative, but when she tries to do so, they tell her to "stay in her lane." When projects hit roadblocks, there is "finger pointing and blame," she says, not the collaboration that is supposed to characterize teamwork. To make matters worse, the culture of the office is sexist—her supervisor believes (and has told her) that a woman's place is in the home. She feels isolated, without a single trusted coworker or a mentor to guide her. Not surprisingly, at that first interview she gave herself a 10-percent chance of remaining at the company, and a 60-percent chance of leaving the industry altogether.

Eve's career was not supposed to turn out this way. Unlike the other two women stayers, she chose her college major with the goal of working for the oil and gas industry. Several of Eve's family members had close ties to the industry, and she graduated from one of the top "oily" schools in the country. She interned at GOG while working on her master's degree, and when she graduated, she received job offers from three other major oil companies.

Despite her negative experience in Australia, Eve was hopeful that her career would take a turn for the better once she and her husband repatriated to the United States. However, a year later, they were back in Houston, and she was still miserable. Once again, her husband's career took precedence. He received a new assignment before returning to the States, but she did not. She spent almost a week in limbo as the company scrambled to find a placement for her. This was a difficult situation to endure, as layoffs were occurring all around her. She suspects that the company had to lay off someone else in order to open up a position for her:

EVE: It was a somewhat turbulent transition.... The problem was that [my husband] had a solid position and I didn't know what I was coming back to. And I didn't find out for sure until I was already back here and I was in a temporary role for three or four days, waiting until they let people go.

CW: So did you replace someone that they let go?

EVE: They say no, in that I took over direct responsibilities of someone who was still there. But there was some shuffling around of responsibilities. It almost seems like it was done to make it seem like I didn't take over someone's role.

CW: That would feel awkward.

EVE: I actually knew a guy on the team who did get let go. And then I came into the team. So it was definitely awkward.... I ran into him in Houston. He has no hard feelings or anything. He was getting close to retirement age anyways. I still feel awkward about it.

At age 31, Eve replaced someone who was "close to retirement age," another indication of the company's systematic preference for younger workers. Further, because the company valued her husband, she got priority for a placement during the downsizing.

However, Eve was not happy with the placement she received. Her new position was drilling onshore wells, an assignment she had explicitly ruled out in her career map. In our previous conversation, she had expressed interest in working in the appraisal phase of drilling, so I asked if she reminded the talent managers about this preference. She replied:

Yep. That's what I told them again, and here I am not in that. I'm not doing that. Yeah, I was definitely asked what I wanted to do. I clearly didn't have much influence on where I was ultimately placed.

Instead of driving her own career as promised, company need (and her husband's career) took precedence. Eve agreed to take the position they found for her, demonstrating the flexibility that GOG values in its employees, and contradicting their commitment to career maps. The only alternative was to lose her job.

Eve expressed many of the same grievances in her new position as she did in her previous one. She did not feel integrated on her team, nor did she feel respected. Even though, at age 31, she was one of the senior members of her 10-person team, she complained that her coworkers did not value her experience and usually ignored her. Unsure how to take control of a bad situation, she tried to find a career mentor, but her attempts failed:

> I tried to reach out to a couple of people, to one guy in particular in a different group. I thought we communicated well, and the jobs he's had are something I'd be interested in. So I was curious and I wanted to see him as a mentor and hear more about how he directed his career. But I felt that we actually don't communicate that well (laughs). The few times I've talked to him about stuff have felt very awkward to me.

Eve blamed herself for her failure to communicate effectively, but the context of layoffs could have compounded her difficulty connecting with this potential mentor. Many respondents told me that, during the layoffs, people began to hoard their dwindling responsibilities, fearful that they would lose their jobs if they did not (appear to) stay busy. Coworkers no longer shared information about increasingly scarce opportunities. That may have been a factor in her coworkers' unwillingness to help her, especially as they were older and probably vulnerable to forced retirement. Besides, Eve was a stranger to them, having relocated to Houston only four months prior. The upshot is that Eve continued to feel isolated at work, ignored by colleagues, and stymied in her professional development.

Surprisingly, during this interview she revealed that she had just received a perfect rating on her performance review for her work in Australia. Based on what she had told me the previous year, I would not have predicted this. However, Eve said, "I busted my butt" on a high-profile assignment and thought she deserved her high rating. I asked if she felt that her perfect score protected her from impending layoffs. Unfortunately, no, she said, because of her recent move:

> In terms of vulnerability now, so I got a "5" last year, but that was a totally different supervisor. I now have a new supervisor who doesn't know me that well. I haven't been in the group that long. I definitely feel vulnerability. They tell you, "It's all based on your performance and your skill set." Yeah, but I know that power plays a huge role in these things. So I don't feel safe by any means. I have no idea how I stack up to people on my team. I don't know. I have not had any meaningful conversations with my supervisor about performance because I haven't been here for that long. And there hasn't been that much going on in the geographic area that I'm assigned to. It's hard to stick out, and I find that very frustrating.

Eve attests to a general problem with the team structure. By the very nature of teamwork, the individual's contribution to the final product is obscured. Members of the team must engage in self-promotion to receive credit and rewards for their personal effort. Women may encounter difficulties when promoting their accomplishments and gaining the credibility of their supervisors and other team members (Williams et al. 2012). In general, women are given disproportionately less credit than men for the success they achieve when they work on teams in male-dominated environments (Heilman and Haynes 2004). In Eve's case, her recent transfer into this group exacerbates this challenge with the team structure. With colleagues who do not respect her, and a supervisor who does not know her, she finds it "hard to stick out" in a positive way.

Eve is concerned about not only her own vulnerability to layoffs, but also her husband's vulnerability. His boss recently left the company, possibly due to forced retirement. Eve's husband was then transferred to a group that was preparing an asset for sale—a sure sign that he along with this entire group would soon be on the chopping block. Without the active patronage that he had previously enjoyed, both were worried that they would be jobless when layoffs were announced the following week.

When I interviewed Eve a year later, I learned that both she and her husband had survived that cut, and two subsequent rounds of layoffs. She had the same position as before, but with yet another new supervisor. She did not know why she was spared at GOG while others were laid off. She thought her perfect score from the previous year could have been a factor. In addition, she felt lucky that the play where she works continued to produce oil, while many other projects languished. Perhaps that is what saved her, she speculated.

During that third interview, Eve seemed depressed and despondent. Although Eve was one of the fortunate ones who survived the layoffs, she told me she was finding it difficult to remain at the company. She desperately wanted to leave, but she felt unable to give up her high salary before lining up another job—and jobs were scarce in the industry at the time. When I asked how she coped with the previous year of downsizing, she sighed and said:

> Oh man. I feel like it's just now that I'm getting over it. It's been two and a half months since the last one. This year was actually harder for me to get past it all, in that I felt a lot of really good people got let go. And that bothered me, the choices that were made. And

then, yeah, I'm exhausted by worrying about this all the time. How do you get through it? I don't know. I became a grouchy sad person. I don't know.

Eve admitted that the layoffs had taken a toll on her mental health. "I thought about reaching out for counseling," she told me. "I feel genuinely depressed. And I feel like that is definitely the result of the layoffs and constantly going through it over and over again."

I asked if the company had made any effort to help the survivors who, like her, were reeling from all the losses. GOG offers an employee assistance program, she told me, that provides five sessions of free counseling. However, she and her coworkers were reluctant to utilize these services. When asked why, she sighed and said,

> The one person I talked to who thought about using it decided not to because the biggest problem is work. Do you really want to go to someone that work is paying for to talk about problems at work?

Eve and her colleagues did not trust the counseling services at GOG to treat their concerns confidentially. After seeing so many talented coworkers lose their jobs, she feared that saying anything negative about the company would make her a target for the next round of layoffs.

Eve's narrative belies many stereotypes about successful women scientists. She is not passionate about her profession, nor is she an advocate of the industry that employs her. She perceives insurmountable barriers to women at GOG, made even worse by the election of President Trump. She told me that she opposes Trump because he boasted of sexually harassing women, but many of her coworkers heartily support him. In this hostile climate, she thought that women could thrive only if they "don't necessarily act like women" by sacrificing their femininity and by never sticking up for themselves or other women.

In one of my final questions, I asked her, "If you had a daughter, or any young women that was asking your advice, would you counsel them to consider a career in this industry?" Her answer is a cautionary tale to advocates for women in STEM:

> Oh, I don't know. The funny thing, I talked to my Dad about this, just with respect to science in general. There is so much push for—oh, we need more STEM people. And you get into these jobs and "I'm not really doing any science most of the time." I don't really think I need the degrees that I have to do what I do on a daily basis.... If I worked at another company where I saw things differently from how I see them where I'm at, I would be more inclined to encourage a daughter to go into it. But where I'm now, I don't think I would.

Eve is one of the three women geoscientists in my sample of 14 who remained after several rounds of layoffs. She is successful by any measure, but she is not a poster child for women in STEM. She is a highly paid professional, but her narrative reveals that she pays a high price to remain in her career. Because of her traumatic experiences in the workplace, she has become a "grouchy sad person" at age 32.

THE GENDERING OF RETENTION

Recruiting women into STEM careers is a priority of government, universities, feminists, and, when the price of oil was high, the oil industry. During the oil boom of the early 2010s, recruitment and retention were constant worries at GOG, as the entire industry feared the upcoming "crew change" when the older generation of professional workers retired. However, once the price of oil fell and companies began laying off their technical staff, these concerns evaporated. Companies slashed their scientific and engineering workforces. GOG cut half of its STEM workers when the price of oil fell.

Lynne, Patricia, and Eve were among the lucky ones who kept their jobs during the downturn. Based on their experiences, what can we conclude about what GOG values in its women scientists? Personality wise, the three could not be more different. Among the three, Lynne is the most enthusiastic: she is an optimistic booster of geoscience, her employer, and the oil and gas industry in general. In contrast, both Patricia and Eve are alienated and feel no loyalty to GOG. Patricia is dreaming about leaving the industry to open a honey-making business. Eve is depressed and feels she is wasting her life at GOG. She doubts she would even encourage young women to enter STEM professions.

Despite these differences, the three women do share a great deal in common. All three received top ratings on their performance reviews. In GOG's forced distribution system, only 10 percent of employees can obtain a perfect score, making them outliers in this particular measure. In addition, all three share similar demographic characteristics: they are young, white, married (in two cases to men who also work at GOG), and childless. Being married to men who work in the industry gave both Patricia and Eve hidden advantages that are not available to others, particularly women of color, who are not only rare at this company, but unlikely to have spouses that occupy high-level positions in the company.

The men in my sample who survived the cuts do not share these characteristics. I followed eight men geoscientists over the years of this study, and four stayed at GOG until the end. (Thus, in my small sample, 50 percent of the men and 20 percent of the women stayed.) These four men are all fathers. All are married, but none has a spouse who is employed by the oil and gas industry (two are married to full-time homemakers). These four men range in age from mid-thirties to mid-forties. Only two of them received top scores on their performance reviews. Three of the four have PhDs (the master's degree is the more typical credential). Three are white, native-born Americans; one is an immigrant from a country where GOG has drilling interests.

These gender differences among the stayers in my sample are remarkable and worth repeating.

- None of the women stayers are parents, but all of the men stayers are.
- All three women received top performance scores, but only two of the four men did.
- All of the women are married to scientists or engineers (in two cases to men who work at GOG); none of the men is married to scientists or engineers (in two cases they have stay-at-home wives).
- Three of the men have PhDs; none of the women does.

These differences indicate to me that GOG's ideal worker is a gendered and racialized norm. Critical race scholar Victor Ray (2019) points out that whiteness is a credential for top salaried positions in corporations, which benefited six of these seven stayers. According to gendered organization theory, companies prefer to hire workers with few distractions outside of work. For men, this means having someone at home looking after domestic matters, including rearing children—which is a racialized family ideal. For women, this means not having children. The company also values technical skills, but the women outperform the men on this measure, suggesting that women must be considered "exceptional" to stay. Having an "exceptional" spouse employed at the company might also enhance women's chances of retention. Three of the four men who stayed have PhDs— a rarity at the company—but they each insisted to me that the credential did not matter in their retention.[7] However, it would not surprise me if the company took educational credentials into account when deciding whom to lay off—something they could not have known for sure.

How significant are these patterns of gender difference? Statistically speaking, they may not matter at all. After all, I am describing seven people out of a sample of 22, which is a fraction of the 88 geoscientists in their cohort at GOG, a company with 20,000 employees. The patterns I detected in my sample may reflect random chance or some undetectable and spurious factors.[8]

The interviews, however, suggest problematic organizational practices at work that could produce these outcomes. Over time, my respondents encountered changes in their work routines that resulted in career shifts that contributed to their survival at the company. Their ability and willingness to accommodate these changes no doubt factored into their retention during the downturn.

During good times, when the price of oil was high, the company's priority was retaining its scientific and engineering workforce. Virtually all the workers I interviewed for this study appeared to benefit from these retention efforts. Some of my respondents were incredulous that GOG even had a retention problem. As one engineer asked me at the beginning of this study, "Why would anyone leave a career in this industry?" When first interviewed, the three women "stayers" highlighted in this chapter were all benefiting from the company's commit- ment to them and their career development. They were working in specialty areas they had chosen; Lynne and Patricia had assumed leadership positions; and Eve was working in a coveted overseas assignment. Once oil prices dropped and layoffs ensued, the company revealed the fragility of these commitments. Career maps were thrown out the window, and only those will- ing to work in areas of company need retained their positions. Lynne changed jobs happily but lost her leadership position; Eve changed jobs reluctantly. Only Patricia remained in her area— fracking—because it was slated for corporate growth.

Many geoscientists I interviewed had to abandon the special- izations they had developed over years of employment. The same shuffling happened to the men I interviewed who stayed. Martin transferred from deep-water exploration to onshore development, a specialty far removed from his expertise. On the plus side, however, he gained a promotion to a leadership posi- tion (at the end of the study, he said he was enjoying his "best year ever" at the company). In another case, Sam learned that his job was moving to New Orleans; he could either move there or lose his job (he moved there with his wife and children the very next week). He, too, gained a welcomed promotion after the move. Much earlier, Sam's wife had abandoned her career as an oil industry geologist and had gone back to school to become a nurse, a career move they thought was more compatible with raising children. (By choosing a female-typed profession, she also reinforced occupational gender segregation with this deci- sion.) Sam considered himself lucky that they were not both employed in the industry during the downturn because his wife could follow his career and remain employable no matter where the company transferred him. For all of these workers, surviving the downturn meant going wherever the company needed them, without regard for their desires or the desires of their family (even though in the end things worked out fine for some).[9]

After the downturn, GOG not only abandoned career maps and their assurance to workers that they could "drive their own careers"; the company abandoned its short-lived commitment to diversity and began promoting inclusion instead. Lynne and Patricia described how the company's leadership—still exclusively white men—were "opening up a conversation" to welcome "all kinds of diversity," not just gender and race. Lynne explained to me that this new "inclusion initiative" was intended to benefit all workers who stayed at the company, including the white men who otherwise might feel excluded by the attention to diversity. This pivoting from diversity to inclusion is occurring throughout the industry today. At a women's leadership conference I attended, the audience learned that white men can embody "cognitive diversity," which in the Orwellian logic of corporate-speak has become more urgent than recruiting and retaining women and minority men.

Lynne enthusiastically embraced this new approach, while Patricia was more skeptical. The only man I interviewed among the stayers who brought up the inclusion initiative was Ben, a Navy veteran with a PhD in geophysics. In his mid-forties, he was one of the older members of his cohort. He felt cautiously optimistic about this new approach to inclusion. He said,

> They've recognized that they haven't been particularly inclusive or open to new ideas, and I think they are a little bit desperate now, and are supposedly going to foster an environment of both safe sharing and inclusiveness. I'm skeptical.

In the past, Ben told me that when he shared his opinions he would suffer "repercussions" if his views did not "fall in line with someone else's thought." His career languished as a result, he thought, as peers with less education and experience received promotions before him. He believes that because the company is now "desperate" to retain their remaining technical staff, they are promising to value critical thinkers like Ben. Ben believes that the inclusion initiative is well intentioned, but he remains skeptical that it can alter an entrenched culture of rewarding "yes men." Pointedly, for Ben, inclusion has nothing to do with gender or race.

The seven geoscientists who stayed at GOG over the course of this study are survivors. As senior scientists, they are now in charge of drilling oil and gas all over the world. They are success stories, but most are not happy. Five told me that the paycheck, and nothing else, keeps them at their desks searching for and extracting fossil fuels around the globe.

The degree of alienation and anxiety experienced by the stayers in my sample is overshadowed by the second group of scientists I will describe—the leavers. In the following chapters, I will describe the experiences of the men and women who left their jobs at the company, and in some instances, abandoned the oil industry entirely.

CHAPTER FOUR

Voluntary Separations

The labor force is constantly in flux. On average, people in the United States work seven years for a single employer before changing jobs (Kalleberg 2018). Two main reasons propel people to move to a new job: either they leave in search of a better job, or their employer forces them out. In general, transferring to a new employer bodes well if workers initiate the job change. Employee-initiated job changes, usually referred to as "voluntary" separations, include transferring for better career opportunities and working conditions, such as improved pay, benefits, and hours. In contrast, employer-initiated job changes, usually referred to as "involuntary" separations, are often associated with wage penalties. These include layoffs due to downsizing or restructuring, and firings due to incompetence or malfeasance (Brand 2015; Cha 2014; Looze 2017).

The distinction between voluntary and involuntary job changes matters a great deal to employers. Companies want to prevent their valued workers from leaving while protecting their right to terminate those they no longer wish to employ. Consequently, they are very interested in learning why talented workers choose to leave. This is the reason why GOG agreed to participate in this research in the first place. When the study began in 2012, oil and gas companies were struggling to retain their scientific workforce. At the time, virtually all of the scientists and engineers in my sample were receiving unsolicited calls from recruiters. I heard a rumor that one company was offering a new truck as a signing bonus. Executives and managers at GOG wanted to figure out what they could do to stop workers from accepting these offers and moving over to their competitors.

Understandably, GOG was not interested in a study of employer-initiated (i.e., "involuntary") separations. When managers gave permission for this study, they did not expect to be laying off half of their technical workforce three years hence. When the downturn happened in the middle of this study, the door swung open to compare workers' experiences of both voluntary and involuntary separations.

The distinction between voluntary and involuntary separations is not always clear-cut. From an individual worker's perspective, there is a gray zone between these two reasons for changing jobs. A worker may feel compelled to leave a cherished job due to poor health or family obligations, for example, so leaving may not feel like a voluntary choice at all. Alternatively, if the work environment is hostile and exploitative, a worker may feel pushed out, even if it is ultimately the worker's decision to move to another job.

In this chapter, I will argue that voluntary and involuntary separations are related in another sense. Companies that lay off workers today may unintentionally provoke large-scale voluntary separations tomorrow. In the case of GOG, the scientists and engineers I interviewed told me that the company's multiple rounds of cuts created a competitive, hostile, and stressful work-place atmosphere, prompting their voluntary departure once industry conditions began to improve. While they may have understood and accepted the inevitability of layoffs when the price of oil fell, from their perspectives, GOG failed to make transparent and fair decisions about whom they retained and whom they let go, creating a toxic environment for the workers who remained.

The business press touts the importance of managing layoffs to ensure that the "survivors"—those who keep their jobs after downsizing—maintain their morale and productivity after the staff reductions are complete (see Brockner 1992 for an over-view). This literature encourages managers to exercise fairness, equality, and transparency in all personnel decisions, and sug-gests that, with the right strategies, surviving employees can learn to accept and even agree with the decision to downsize the company. These writers blame low employee morale among survivors on the improper management of layoffs, not on layoffs per se.

This management literature is based on the assumption that workers no longer expect job security, but anticipate being laid off at some point in their careers. This is also the conclusion of cultural sociologist Allison Pugh, who sensitively portrayed the lives of laid-off workers in her aptly named book *The Tumbleweed Society* (2015). Gone are the days when a sense of loyalty bound workers and employers together, she found. Pugh identified a new social contract for our precarious times, which she labeled the "one-way honor system," in which workers pledge their hard work and loyalty to their employer in return for a paycheck. What makes this system "one-way" is that workers maintain high standards for their own work ethic while expecting their employers to prioritize corporate interest and profit ahead of their well-being. Workers may long for jobs that enrich their knowledge and skills, but they no longer expect this, according to Pugh. Instead, workers perceive career development opportunities as an extra benefit of employment, available only to the select few who have proven themselves to be top performers. In this way, workers shoulder the blame for their failure as well as take credit for their success, absolving their employers of any responsibility for how their careers unfold.

This chapter highlights the institutionalized practices that generate the "one-way honor system" by delving into the personal narratives of the eight scientists in my sample who voluntarily left GOG to work for another oil company. Their choices to leave GOG reveal what workers expect from their employers and what they are willing to give in return. They also show how periods of corporate downsizing can result in rewriting the employment contract, as workers reassess their loyalty to their employers.

Four of these "voluntary separations" occurred during the boom, before the fall in oil prices and before the layoffs started. These scientists liked their jobs at GOG, and described mostly pull factors prompting their decisions to leave. For them, other oil companies offered a better "fit" for their skills and personal lifestyle preferences. The other four scientists in this group transferred to another oil company after the bust, when oil prices began to rise again and the labor market started to recover. They described mainly push factors in their decision to leave. They left to escape deteriorating working conditions resulting from multiple rounds of layoffs. Their narratives provide cautionary lessons to corporate downsizers, showing how mass layoffs can promote voluntary employee attrition— precisely what companies claim that they want to prevent.

The eight geoscientists in my study who voluntarily left GOG for jobs in other oil companies include five women and three men. All but two are US-born white people. About half are married, but none has children. For the labor force as a whole, workers without children are the group most likely to voluntarily change jobs—a pattern that holds true for both men and women (Hollister and Smith 2014). Married parents may feel unable to switch jobs due to their family's reliance on their job benefits, including health care, a predicament sometimes called "job lock" (Bansak and Raphael 2008).[1] Thus, it is not surprising that in my study all of those who voluntarily left GOG to work for other oil companies were childless at the time they left.

PULL FACTORS: VOLUNTARY SEPARATIONS DURING THE BOOM

Of the four geoscientists who left GOG during the boom in oil prices, two were women and two were men. In three of these cases, their reasons were location-related: they jumped at the chance to leave Houston. Two were self-described small-town people who pursued opportunities to work for companies located in rural parts of the country. The third left because he loathed everything about Houston, aside from his job. Steven spent a good part of every interview cataloging his many reasons for despising the city, including its politics, people, culture, climate, and traffic.

The fourth geoscientist in my sample who transferred to a new oil company during the boom remained in Houston and joined a European-based firm (where her geologist husband also worked). Elena transferred the second year of the study. When I interviewed her the following year, she told me that her reasons for leaving GOG were mostly career-related. A promised position in exploration never materialized at GOG; instead, the company assigned her to a development role that did not make use of her seismic expertise. She told me that her new firm offered a much more challenging and stimulating work environment, which is the main reason she transferred.

Elena was pregnant the first time I interviewed her. When she had her baby the following year, she took four months' paid maternity leave—a more generous arrangement than offered at GOG. At GOG, new mothers could take six weeks of paid disability leave and up to six weeks of unpaid leave; any additional accommodation required their supervisors' permission. Fathers were allowed to take one day of paid paternity leave at GOG, but only if their supervisor allowed it. Elena's new company, in contrast, offered eight weeks' paid leave to new mothers, followed by either four weeks of part-time work at 100-percent pay, or another four weeks off for 60-percent pay. New fathers received four weeks of paid leave to use any time within the first six months of the baby's birth or adoption. These arrangements did not depend on supervisory approval. Elena told me that she was not aware of her new company's maternity benefits when she decided to leave GOG—she was not pregnant at the time— although she definitely appreciated them now.

However, Elena did indicate that diversity was a "very important" factor in her decision to transfer to a new firm. She rated this factor highly on the survey she filled out, which asked her to identify her top reasons for leaving from a list of options we provided. The other "very important" factors she identified in her decision were "friendlier environment," "more opportunities," and "more interesting projects." When I asked her to elaborate on why she selected "diversity" from the list of reasons she left GOG, she drew links between all four of these factors:

> I think because I'm a younger person and a woman maybe you hope to see more people that are sort of like you in your workplace. A lot of places, especially in like US onshore, it's still kind of that old boys' club, especially in some of these older oil companies like GOG. It's mostly middle-aged men. I really never had a career role model.

Elena told me that at GOG, where so few women occupy positions of power and authority, "it just makes you feel like it can't be done, like nobody has done this." In contrast, her new company kindled her ambitions for management. "It's a lot more international and a little bit more egalitarian," she told me. "I've had a lot more women supervisors here than I ever had at GOG." At the time she left, she thought that a company that valued diversity would also be a "friendlier" environment—not an old boys' club like GOG—that would provide her with more stimulating career development opportunities.

Unfortunately, the following year layoffs hit her new company and Elena lost her women supervisors. Only two women in senior management remained at the company, she told me. Although the employee cuts were not as drastic as those at GOG, Elena regretfully acknowledged, "There aren't a lot of women doing this work." Including her, only three women worked on the entire floor of her open-plan downtown office building.

Although dissatisfied with the diversity at her new workplace, Elena remained satisfied with her job because it stimulated her scientific curiosity. "It's never dull," she explained. "There is always something new to do. A new data set or a new kind of data, or a new tool for doing something." Elena and her husband were enjoying their work and hoping to survive the downturn.

Elena's voluntary separation from GOG was a success, but changing jobs always carries risks. The new company may not live up to expectations, and leaving can increase workers' vulnerability to layoffs if the new company decides to eliminate those who are "last hired." This fate befell the three people in my sample who left GOG for geographical reasons, mentioned earlier. All three left their new jobs within a year of leaving GOG (but subsequently found industry employment elsewhere). In contrast, Elena felt relatively secure at her new company. As suggested in the previous chapter, having a husband in the same company might have provided her some extra protection from layoffs.

Despite the insecurity experienced by the others in this group, all four who left GOG before the crash in oil prices made what they considered positive choices: they pursued better opportunities at a different oil company that suited their geographical or professional preferences. None said anything negative about GOG, at least not at the time they left.[2] Instead, they told me that they had left GOG because they preferred to live and work elsewhere.

Could GOG have done anything to prevent these voluntary departures? Providing the option to telecommute to those who left for geographical reasons might have convinced them to stay, as telecommuting in general promotes retention and enhances productivity (Glass and Noonan 2016). Increased diversity might also help to retain valued employees: Elena left GOG desiring more women role models and a friendlier environment, which she found in her new European-based firm (although their numbers dwindled after the downturn). Doing a better job matching jobs with skill sets also might have prevented Elena from leaving GOG, a theme that emerges in the next section as well. For a multinational corporation like GOG, implementing these changes would be difficult but not impossible, and might have stemmed the voluntary attrition of these valued scientists. For whatever reason, GOG did not act to retain these valued employees, betraying the limits of their employment contract.

PUSH FACTORS: VOLUNTARY SEPARATIONS AFTER THE BUST

In contrast to the positive accounts of those who left before the downturn in oil prices, the four geoscientists in my study who transferred after the downturn when the market began to recover described their decisions to leave in starkly negative terms.[3] This group includes three women and one man. When they left to join another firm, they did so to flee a work environment they considered intolerable. Three main reasons emerged from their accounts about leaving GOG: the lack of meaningful work, toxic relationships in the office, and stress occasioned by the layoffs. Each of these draws a connection between their voluntary decision to leave and the involuntary separations that the company imposed on their coworkers. Each also reveals the organizational practices producing the "one-way honor system" between workers and employers.

Lack of Meaningful Work

As we saw in chapter 2, many of the geoscientists in my study joined the oil and gas industry for the prospect of high pay and exciting work. During the downturn, however, the availability of the latter diminished. The decline in oil prices meant that the primary work of geoscientists—finding and exploiting oil and gas reserves in the earth—began to dry up. Although some did experience increased workloads after the crash, others saw their job duties diminish, making their workdays almost unbearably bleak.

For example, Kimberly found herself without challenging work to do. She said:

> There is just not enough work. I kind of describe it as detention. You know, like when you have detention you're supposed to write your spelling words or copy the dictionary or something.

When the price of oil was high, Kimberly was given opportunities to manage projects, but when the business faltered, she was moved to a new group and given very little to occupy herself. Even though she survived multiple rounds of layoffs, she told me that she personally experienced downward mobility over the course of her tenure at the firm. She said,

> I went from putting together and executing projects that are like 250 million-dollar-plus wells with a lot of responsibilities and oversight, to doing nothing. I feel like I've taken a step back in my career. I think [my current assignment] is a safe harbor, but if there wasn't a downturn, I wouldn't have wanted to come to a group [like this one] that doesn't have really operative activity.

No longer drilling for oil, Kimberly was miserable. "I'm starting to feel like [the movie] *Office Space* with the stapler," she told me. "Just don't take my stapler away from me and I'll be fine."

When the price of oil started to rise and hiring opportunities began to open up at other firms, Kimberly decided to leave GOG and accepted a job (and a promotion) at another oil company. Although she was given many more responsibilities at her new job, her workload increased too much, she told me. She experienced overwork, which required being on-call at all hours. Owing to the constant stress, she lasted less than a year at that company before transferring to yet another oil firm, hopeful of finding a better work-life balance.

Gloria shared a similar experience to Kimberly, as GOG shuffled her from a responsible position to a lower-level administrative role during the downturn. Sighing, she compared the company climate at GOG before and after the fall in oil prices:

> I just like new ideas, I like people who have ideas and want to execute. I feel that was happening here when times were good. I think people are just wasted now. Just completely emotionally and mentally wasted. Intellectually wasted.

Gloria told me that all of her colleagues planned to leave GOG as soon as they received their long-delayed bonus checks. The company stopped paying bonuses during the crash, so these workers were postponing their departures until oil prices began to recover and their promised bonuses finally materialized.

Throughout his tenure at GOG, William struggled to find a position that utilized his dual training in geology and business. Having a unique combination of a PhD in geology and an MBA degree, William desperately wanted the opportunity to develop both of these skills, an aspiration that GOG seemed unwilling to enable. In our first interview, he said:

> I've told the company repeatedly for years that I'd like to be more involved in the commercial and business aspects. I'd like to be able to pull on my business degree as much as I pull on my geology degree, and they just don't [let me]. I don't know if it's lack of creativity or just a total lack of sensitivity and listening to what people say they want to do with their careers. But they simply don't have a way to make that happen.

William felt his talents atrophying at work. He complained that his job was making him dumber, not smarter—a classic form of alienation. He speculated that the company's inability or unwillingness to utilize his multiple abilities had to do with their forced ranking system. He perceived that only those identified by the company as "high potentials" received career development opportunities, and he was not one of those (for reasons I will explain later in the chapter). He said,

> I think the thing at GOG, they identify a small group of people that they really want to focus on. Ten percent, 15 percent, I don't know what it is. But they will bend over backwards; they will do whatever it takes to develop that 10–15 percent. If you're not in that category, you are simply not someone that the company wants to spend resources and time and attention on. You are 100 percent expendable. "I really don't care if you leave, if you stay for a while, whatever you do." If you're not identified as being one of those high-potential people, it's like you're not on any list for any career development.

William thought he found what he was looking for in a smaller company. Six months after transferring there, he assumed a supervisory position with a three-million-dollar budget to manage. "It's great," he told me. "It's a complete 180 from the track that I was on when I moved." I told him that he sounded much happier than in previous years. Agreeing with me, he said, "I just have a sense of purpose working here now. I feel like I belong to the team and I'm treated well by management, and I love it." His is a textbook example of a voluntary job change resulting in upward mobility. Although he did not like everything about his new job—he had to lay off workers in the first year of his employment (discussed later)—his new job was more rewarding financially and intellectually compared to his position at GOG. In his case, his voluntary job change in pursuit of meaningful work appeared to pay off.

The lure of voluntary separation in pursuit of meaningful work came up often in my interviews, when times were good (as in the case of Elena), when times were bad (as in the cases of Kimberly and Gloria), and in both scenarios (in the case of William). Many of the scientists and engineers I interviewed saw addressing the dearth of meaningful work as key to solving the retention problem in the industry, a perspective widely shared in the management literature. One of my respondents sent me an article from *Fortune* magazine that made this precise point. Titled "10 Reasons Your Top Talent Will Leave You" (Myatt 2012), the author asserts that retaining employees depends on cultivating their passion, intellect, creativity, and skills. While all of these factors mattered to the people who left, notably missing from the *Fortune* list is the sense of security or community, or the sense of "belonging to a team," as William put it, quoted earlier. Meaningful work, like meaningful anything, is a product of social interactions based on mutual respect and recognition. This cultural dimension is lacking in the management literature, which promotes radical individualism instead, in which each talented worker is imagined on an existential quest to find personal fulfillment. That this fulfillment may be unattainable in the context of widespread employment insecurity is rarely addressed.

Income is also curiously missing from the *Fortune* list of what top talent desires. My respondents told me that their high salaries compensated for the lack of meaningful work, but only to a point. They were willing to move laterally for the chance at more interesting and fulfilling work, which was in short supply at GOG after the downturn.

Toxic Relationships

All four of those who voluntarily left GOG after the downturn described increasingly toxic relationships at the company. They complained about mismanagement, lack of recognition, and constant threats, humiliation, and bullying. When oil prices fell, Leticia was transferred to a new group comprising mostly young people. All were competing to keep their jobs, contributing to a hostile work environment. To illustrate her miserable situation, Leticia gave these examples: "You are doing a presentation, and they tell you 'That's wrong; you shouldn't have done it that way.' Or someone takes your work and presents it before you are finished—like it's their own. Backstabbing, things like that." Her new supervisor, who was a woman, meted out "harsh criticism" without giving any direction on improvement. Although they were the only two women on the team, Leticia suspected that her supervisor targeted her for abusive treatment to avoid giving men the impression that she was biased in favor of women.

Gloria, too, suffered from mistreatment by her supervisors, describing them as passive-aggressive and duplicitous. One boss asked her to give an informal presentation about her work, without mentioning that its purpose was to evaluate her for a supervisory position. Unaware of the importance of her presentation, she spent little time preparing and failed to impress the top managers who attended, resulting in her transfer to a lower-level position.

Gloria described her new job as dismal and boring. Even worse, the position compromised her professional ethics. Her new boss demanded she approve his pet projects even when they lacked scientific evidence to support them:

> He knew the answer, and he wanted you to show him that was the answer, no matter what. And that is not how I work.... I realized there were things that weren't working with the project, and that was pissing him off. That's science, I'm sorry. And I'm first and foremost a scientist, and I honor the data. And I was not happy.

Gloria attributed the toxic environment she experienced to the downturn, which was making everyone at the firm feel threatened with job loss. Under the circumstances, scientists faced pressure to approve all management decisions no matter how unreasonable they were, becoming proverbial "yes men." "You're in the tribe or you're not," she told me. "It was ridiculous." As soon as she received her bonus, she left GOG for another Houston-based oil company.

The perceived arbitrariness of annual performance reviews exacerbated the toxic environment of the firm, according to the four who moved to another firm after the recovery began. They considered the process of evaluation at GOG as deeply unfair. Kimberly, who was stuck in "detention," had never received an "outstanding" performance rating during her tenure at the company, and she was at a loss to understand why. The process was not transparent, she told me, and the supervisors charged with evaluating her performance did not take their responsibility seriously. "Probably the worst that I ever had, someone put in my performance review that I'm a nice girl," she sighed.

As mentioned previously, before her exile into detention, Kimberly was in charge of drilling a multimillion dollar well, a job at which she excelled, she told me. Yet the perfect "five" eluded her. When she inquired why, her supervisor explained that Kimberly "performed like a five" but because they could assign only two fives, they ranked her a four instead. I asked her what that conversation was like for her and she said:

> It's incredibly demotivating.... It's like being in school and being told we only have two 100s that we can give out. "I know that you got the questions right, but you are the third paper we picked up, so you got a B." (laughs) It's like, why even try? I think that is the message that that sort of system signals. We talk about that it's about performance and accountability, but when you have arbitrary cut-offs that are not on the merits of what that person's done, then it's not performance or accountability, it's [pause] happenstance? I don't know.

Adding to her frustration, Kimberly claimed that the only time she did receive accolades in her career was for organizing an office party. This happened when the price of oil was still high. We were talking about her performance evaluation, and she told me this story:

> We had this big meeting last week and I get assigned to set up the food for it. And I'm like, whatever, I'll set up the food. I get more compliments about how great of a party I can put on than the work that I do. And I'm, "What is this?" ... It's just not fulfilling when you get more compliments on the type of party you can throw than the technical work you can do. Somebody saying, like "Hey, thanks for staying late and getting me those numbers, I know it was a sacrifice." Instead it's like "Great party! You picked a good restaurant." I mean, anyone with access to Google can do that. It doesn't take any specialized training. I just typed in restaurant reviews.

Kimberly longed to be acknowledged for her technical abilities. Instead, she told me, "the male engineer gets to lead the project, and I'm working my tail off, getting numbers, doing things, but I get the attention for planning the restaurant." During her tenure at GOG, she never felt acknowledged or appreciated for her scientific skills.

Even though she was deeply dissatisfied with the performance evaluation system, Kimberly's supervisors and the talent management team encouraged her career development during the oil boom. Since the downturn, however, it was "every man for himself" as the managers worried about protecting their own jobs. She used the term "man" intentionally, perceiving a gender dimension to her plight. Sexism was clearly behind her "nice girl" evaluation and her party planning accolades, she thought. Kimberly also perceived that managers were targeting women for layoffs, despite the company's reassurance otherwise. "When I look around the petro-technical side, it kinda looks like the oil field of yesteryear," she said, exaggerating her soft Southern drawl.

However, Kimberly found it impossible to complain about sexism—except to me. She had joined the women's network at GOG, she told me, in order to improve the situation for women at the company. Unfortunately, she felt she could not articulate her concerns at their meetings, even though—or perhaps because—the company's personnel manager attended its functions. In the context of the downturn, Kimberly perceived that complaining was too risky. This view was underscored for her by the CEO, who told employees at a town hall meeting that instead of complaining, they should be happy that they still had a job. In this context, Kimberly feared that complaining would make her vulnerable to being laid off, and might cause her to develop a negative reputation that would follow her to her next firm. In the small world of petroleum geology, she feared she would become a pariah. This is why Kimberly hoped to relay her concerns about the company through me, believing that executives at the company would listen to me and heed my recommendations for change. I could complain about sexism on her behalf but she could not do so without suffering irreparable damage to her professional career.

William also suffered from what he perceived to be an unfair evaluation, but he did complain—and suffered consequences for doing so. While stationed abroad in GOG's Canada office, his supervisor rated him a "two," a subpar rating that would normally result in him being fired or laid off. But William was spared this fate because, on the same day he received the low rating, he also received news of his promotion. All of this happened exactly one week before he was scheduled to return to GOG headquarters in Houston (GOG ended all international assignments during the downturn). In fact, William learned about both his promotion and his low rating at his going-away party in Canada.

Back in Houston at GOG headquarters months afterward, William was still seething about his "two" rating even though he was spared from being laid off. I asked William to speculate why he got such a low rating even though his performance merited a promotion. He attributed it to his former supervisor in Canada, who harbored a grudge against him for two reasons. First, William had previously stood up for a woman employee whom the supervisor had threatened (William encouraged the woman to file a formal complaint against the supervisor). The low rating was a form of retaliation, he thought, for defending his female coworker.

Second, William speculated that his supervisor gave him a low rating to protect the Canadian members of his team, who remained in that office. He explained:

> He knew that they were going to be laying people off in the first quarter of 2015. I would have been [back in Houston] by then. I would have no longer been his concern. There are a certain number of these two's that they have to give out. Rather than give it to one

of his people, who actually deserved it, and have to lay them off, he gave it to me because I would be long gone by the time they had layoffs.

In the forced rating system at GOG, a certain percentage of workers must be assigned low scores. William thought that he was assigned the low score because his supervisor was trying to protect the other members of the team, who would be staying in Canada after William left. In other words, by assigning William the low score, the supervisor could protect the rest of his group from layoffs down the line.

Back in Houston, William's new supervisor told William to simply forget about the "two" rating—which he dismissed as "political"—and tried to reassure him that it would not affect his future at GOG. However, once again, William said he was assigned to a "dismal" job that did not draw on his multiple talents. He immediately started searching for a position at a different oil company. Once he found a new position and announced his resignation from GOG, a vice president at GOG invited William to an exit interview. When asked why he was leaving, William explained it was mostly due to the low rating, something the VP claimed was news to him. Exasperated, William exclaimed:

> The shocking thing about it, when I got back to the US, I talked to K., my manager, about this, a lot. He assured me that he had discussed it with the VP, and that the VP's perspective was that it was all politics, not to worry, there wouldn't be any damage to my career. Bite the bullet and move on. But K. had never talked to the VP about it! The VP was shocked to hear about it; he had never heard about it. He had no idea that this happened. I was so disgusted at that point. Whatever, that's fine. That makes sense that you never heard about it. K. lied to me. That's the way it works there.

From William's perspective, his unfair evaluation poisoned his experience at the company. Complaining did not result in redress at GOG, but only in retaliation and lies.

That workers who experience toxic relationships choose to leave is not surprising. Throughout the labor market, "exit" is a popular choice among workers unhappy with their situation (Hirschman 1970). Perhaps surprising is that the toxicity experienced by these voluntary departures appears to be the direct result of the company's actions during the downturn. As GOG cut its professional workforce, it perhaps unwittingly created the very conditions that led to the voluntary departures of its valued employees once the job market improved.

Each of these leavers attests to a new atmosphere at the company post layoffs, characterized by competition, backstabbing, and deceit. Rolling layoffs bred job hoarding as workers feared inactivity might put them on the list of the next to go. To garner their supervisors' approval, workers took credit for their team members' efforts and learned never to criticize their supervisors, even when they thought they were wrong-headed. These workers learned not to trust any official at the company to tell them the truth. The climate of insecurity at the company transformed working relationships into dysfunctional chaos.

These interviews suggest a gendered dimension to this toxic environment at the firm. As the industry contracted and layoffs unfolded, Kimberly perceived that women either lost their jobs or were transferred to low-level positions, which was also Gloria's and Leticia's experience. These three women saw exit—and not voice—as their only alternative to escape the toxic relationships they experienced at GOG (Hirschman 1970). Not that only women are affected. William stood up for a woman colleague (and himself), and he faced retaliation for doing so, suggesting that allies also may suffer for complaining about sexism. In a context where multiple rounds of layoffs are unfolding, any expression of discontent—let alone resistance—may be grounds for dismissal. These four scientists left as soon as they lined up a new job, in an effort to obtain the respect and recognition they felt they deserved.

Stress

The four people who voluntarily left to go work for another oil company after the downturn survived several rounds of layoffs at GOG. In this sense, they were fortunate compared to their laid-off coworkers. However, witnessing others lose their jobs contributed to their decision to voluntarily leave the firm once the labor market improved.

The involuntary departure of many valued colleagues was intensely stressful to them. Leticia, for example, told me that she took several months off after leaving her job at GOG before starting work at her new company in order to recover from the stress. She said,

> I got really tired.... People being laid off, a lot of sadness, a lot of competition—put a lot of stress on me. I needed to take some time off to relax and be able to enjoy everything again.

Layoffs were a constant threat over the course of two years. In each cycle, it seemed, the stress level increased. She said,

> They announced layoffs in February to happen in April. So two months of waiting to see if you are going to be laid off. And then in July, they announced layoffs would happen in October. So that was even worse, because it was a longer wait. And that wait makes the environment very heavy and difficult.

William concurred about the prolonged misery of waiting for layoffs, noting that GOG "has done it in bites" over the course of the downturn. He reasoned, "It's probably better to do it earlier and cut deeper than to string this out.... It puts people through a lot of stress."

Those who survived the layoffs were devastated as wave after wave of their coworkers and friends were forced out of the company. Kimberly described the process of layoffs as inhumane and humiliating, describing how her coworkers "went from an employee to a trespasser in a moment's notice." In some instances, managers told groups of coworkers to report to either Room A or Room B; everyone in Room B was laid off and immediately escorted out of the building by security guards. In other instances, workers arriving for work in the morning were stopped at the front reception desk and told to return to the parking garage because their jobs no longer existed. They did not get the chance to say goodbye to their coworkers or to collect personal items from their offices. One of my respondents said her supervisor instructed everyone on her team to work out of boxes in order to make it easy for him to mail them their personal effects in the event they were laid off.

Gloria told me that this constant stress was devastating for those who left as well as for those who stayed behind:

> I stopped thinking about it after a while because you just can't. It would get depressing. You'd block it out and just do your work. It's tough! It's tougher than graduate school. I would take it any day, sitting in front of my committee and have them hammer me with questions. Going through 3 to 4 layoffs over the last two years? Watch your friends get laid off for reasons that were just like, "Oh, there is just not a position," and they're questioning their self-worth as they're walking out to the parking garage. It's horrible. It's

horrible because we like to define, I think, in our society, people by what they do. And it's devastating when they lose their jobs.

A sense that the layoff decisions were arbitrary contributed to the stress of the survivor guilt they experienced for keeping their jobs.

The company did little to assuage their stress. When asked at a company town hall how layoff decisions were made, the CEO reportedly told the mass gathering "the best player plays." This statement felt like an insult to Kimberly, who pointed out to me that GOG carefully vetted all employees before they were hired:

> They keep referring to this "best player plays." I think it's unnerving when you've got the CEO saying something like that when they've hired every single person here. No one just showed up here and was like, "I demand to have a job at GOG." This company has hired every single one of these people. And there's not that leveling [with us] to say, "Hey we're in a hard time, we've got to make some hard decisions, as much as it depends on me as well as the leader of the company, I'm going to try to find something for everyone. But understand that we will have to make some hard decisions that will impact people's jobs." Rather than "best player plays" [said glibly]. What is that? This isn't like a general population of random people. These are people that you've made a commitment to for a job. They've spent their time coming to your organization and providing for it. It seems like they treat it more like fifth grade dodge ball.... I don't see the tact, the dignity, the human element that says, "at the end of the day, this company would not be anything were it not for the people who show up every day to make this company happen."

Kimberly expected her employer to treat workers with integrity and dignity—values that were inscribed in the corporate mission statement. However, these values were nowhere in evidence during the time of industry contraction, revealing in stark form the one-way honor system, and motivating Kimberly to leave GOG once industry hiring picked up again.

The flagrant imbalance of the one-way honor system revealed during the downturn caused William to rethink his own feelings about company loyalty. After his experience at GOG, he will never again expect anything from his employer, he told me. Speaking from the vantage point of working at his new oil company, he said,

> It's kind of weird. Your first company that you think about leaving, it's an adulterous feeling almost. You're ashamed of it. I'm just completely over that. I'm loyal in the sense that I work hard for the company. I'm here sometimes 10 hours a day. And I work hard for our success as a group. But I'm never going to let myself get back to a point where I define my identity as a person based on my employment, where I'm an employee. I'm just not. I did that with GOG and it was an awful experience. And I don't want to put myself in that position again.... Even here I don't have any false illusion that that kind of loyalty is going to be returned in equal measure. I just know how it works. I'll be loyal in the sense that I will work hard for whoever I'm working for. But I'm also going to be open to other opportunities.

Now that he is fully aware of the one-way honor system, William resolves to restrain his feelings of identification and loyalty to his employer. He promises to work hard, but only for a paycheck. An employer is not worthy of his emotional investment.

As it turned out, soon after he arrived at his new company, William had to lay off members of his team, although he was unaware of this at the time. Right before he left for vacation, he submitted performance evaluations for the eight people on his team; when he returned, he found out to his horror that four had been laid off based on his reviews. "It was, my first time being in a leadership role like that," he told me. "It's difficult to deal with." However, he accepts that being in a leadership role means laying people off. The harsh lesson learned is that every single employee is disposable and should always be pursuing other opportunities. According to the one-way honor system, it is foolish to expect anything else.

EYES WIDE OPEN

Oil companies do not usually invite outsiders to study their employment practices. However, GOG made an exception out of heightened concerns about attrition during an oil boom. With the "great crew change" on the horizon, they opened their doors to this study, hoping for advice on how to stop voluntary attrition among their highly valued scientists.

Companies like GOG may be sincere in their expressed desire to retain their best workers. However, this desire is always conditional. As they fret about attrition, companies also demand the right to downsize their workforce at any time for any reason. GOG sought loyalty from workers but refused to reciprocate except through a paycheck (albeit a very lucrative one). The narratives of the eight voluntary leavers highlighted in this chapter reveal the costs of this one-way honor system.

GOG did little to retain its prized employees before the downturn. When technical skills were in high demand, scientists employed at GOG shopped their talents around to other companies in search of a better fit. For the four who left GOG during the boom, this meant pursuing career development opportunities and lifestyle choices that GOG was not able or not willing to accommodate.

After the downturn, when the job market began to recover, four of the scientists in my sample left in order to escape deteriorating working conditions that they blamed on management decisions at GOG during the downturn. Importantly, they did not question downsizing per se. Downsizing is considered an inevitable part of this industry, and every oil firm is expected to periodically lay off employees. However, these leavers held out hope that a different company would treat them with respect and dignity.

Through its multiple rounds of layoffs, GOG aggravated alienation, toxic relationships, and stress for these workers. While these four employees avoided the personal humiliation suffered by those who were laid off, watching wave after wave of their colleagues being dismissed without dignity or respect convinced them that the company was no longer worthy of their labor and commitment. Their perceptions that the company's actions were arbitrary, politically motivated, or sexist convinced them to take their skills and talents elsewhere once labor market conditions improved.

Companies wishing to retain their talented workers would do well to heed this lesson: Widespread use of involuntary separations can provoke voluntary separations in the future. Convincing valued employees to remain requires that firms treat their workforce with dignity and respect—qualities in short supply during the multiple staff reductions.

Is it possible to do both? Can companies both treat workers humanely and engage in downsizing on demand? The management literature suggests that the answer is yes, but I am not so sure. Downsizing by its very nature is dehumanizing. When layoffs become a routine part of a business cycle, as they are in the oil industry, employees learn that they are disposable, regardless of their performance. This is the opposite of being treated as a human being deserving of dignity. It is disingenuous for companies to complain about voluntary separations without addressing their own commitment problem first.

Not only are they inhumane; layoffs can exacerbate inequality inside organizations insofar as they target women and minority men (Byron 2010). However, this is not the inevitable result of downsizing, according to economic sociologist Alexandra Kalev (2014, 2020). In her view, layoffs may be "a necessary evil," but they do not necessarily reinforce white male domination inside firms. Her research found that companies taking a performance-based approach to downsizing can avoid exacerbating the inequalities of gender and race in the aftermath of layoffs.

Unfortunately, my interviews tell a different story. At GOG, although performance was a criterion for retention (along with skill set), it was widely understood by my respondents to be a "political" assessment, based on supervisory discretion and bias. The four scientists who voluntarily left after the downturn complained about the unfairness built into the performance review system, which was perceived to penalize those who spoke out against corporate practices, or who deviated from the company's expectations in any way. Nevertheless, these four held out hope that a different oil company would treat them equitably, fairly, and with transparency. Consistent with the management literature, they believed the problem was with GOG, not with downsizing.

Even though layoffs may be considered a "necessary evil" in the oil industry, the geoscientists in my study seemed utterly unprepared for the magnitude of the losses they witnessed at GOG, and taken aback by the callousness exhibited by the company's leadership during the staff cuts. Why did they not see this coming? Some of the geologists I interviewed had family members in the business, including some who had lost their jobs during previous industry downturns. Why weren't they better prepared for what happened?

For his part, Steven, who left the company during the boom, accused the company of intentionally misleading him and other members of his cohort. Recruiters promised them that conditions had changed in the industry, and that their jobs would be secure. Steven thought that he and his coworkers were easily hoodwinked because they were young scientists without "a lot of background about business and the bottom line," he told me. Speaking from his vantage point working at a different oil and gas company, he described GOG's duplicity:

> The dumb bastards who back in 2009 or '10 who were hiring like crazy, worried about the gap, were telling everybody this is a great career: "You've got a bright future ahead of you, I'm telling you, there is a big crew change going on here." They should have known better! And if they didn't, they were either dumb or they were misleading to a lot of us young folks, to the point of being outright lies.

GOG lost all credibility with Steven. Although he left the company during the boom on positive terms, he was appalled by the company's subsequent treatment of his former coworkers during the bust, whose experiences he followed from afar. The duplicity that GOG showed toward these workers characterizes all of their business dealings, he asserted. Managers who lie to employees will lie to anyone:

> These guys will say anything. They will say anything to get done what they need to get done, [including] bending the facts for oil

and gas.... And you know the shit that they're spouting externally to the public—oh my god!

Steven was referring to the "greenwashing" publicity campaigns undertaken by the company to shore up support for the fossil fuel industry. There was not a shred of truth in anything the company said. He regretted that he ever believed their lies.

Steven took great delight in being recorded during our interview. "I love the idea of you listening, or anyone listening back to this tape and hearing me call them DUMB BASTARDS because I think they potentially screwed up a lot of people," he exclaimed. He and his coworkers learned the hard way that industry representatives lie. However, if they were gullible then, now their eyes are wide open. "It was a rude awakening," he admitted.

I asked Steven if he feels loyalty to his current company. He said,

> In some ways, yeah. But [pause]. It depends on your definition of loyalty. Certainly professionally, I'm looking ahead to when I'm going to walk. So I guess I don't feel loyalty in that way.... If your definition of loyalty is something about willing to bend the truth, or spin a really good story about something that isn't there, then I don't think I feel that kind of loyalty. I feel loyalty to our organization in that my impression is they do things better. They aren't a shady group. They do things a little better than some of their peers. And they've treated me pretty well. So I'm loyal to them that way.

Steven's loyalty extends only to the truth. He works on the condition that his employer will not lie to him or ask him to "bend the truth." In the end, he realizes that this may be too much to ask of an oil company. Although he loves his high salary, he admits his days in the industry are numbered. The one-way honor system is not enough for him. He's looking ahead to when he's "going to walk."

The next chapter looks at the experience of those who did walk away from the oil industry. All casualties of downsizing, they left oil and gas behind them when they left GOG.

CHAPTER FIVE

Corporate Downsizing

The decline in oil prices in 2015 prompted oil and gas companies to cut massive numbers of jobs. In the industry as a whole, over a quarter of a million jobs were lost that year, with cuts concentrated in the upstream areas of the business (where most oil industry scientists are employed).[1] Between 2014 and 2016, GOG cut one-third of its global workforce and eliminated 50 percent of its scientific and engineering positions, as they sold assets and cut entire divisions of the company.

Among the entire cohort we studied at GOG, a third of the men and half of the women left the company during the period of downsizing. In my small interview sample of 22 geoscientists, four of the eight men left (50 percent) and 11 of the 14 women left (80 percent). Of the 11 women who left, five were the victims of downsizing.[2] One was force-retired, two volunteered for severance (which I will explain shortly), and two were laid off. This chapter highlights four of these cases, all of whom were mothers of young children.[3]

Targeting mothers for layoffs is a particularly brutal expression of the "motherhood penalty," the well-documented bias experienced by mothers in the workplace. Sociologist Shelley Correll and her colleagues have demonstrated that mothers experience discrimination in hiring and promotions. They also receive lower pay compared to fathers, and also compared to men and women who do not have children (Correll, Benard, and Paik 2007). This chapter argues that, in addition to those forms of economic discrimination, mothers experience heightened vulnerability to layoffs (see also Byron 2010).

In their enthusiasm to bring more women into STEM occupations, the educational and policy establishments rarely mention the possibility that scientific careers can be precarious (Teitelbaum 2014). However, scientists' vulnerability to layoffs may be key to understanding why women leave STEM occupations more often than men do (Glass et al. 2013; Hunt 2016). As employers churn their labor force, either in response to economic crisis or to appease investors, they are presented with opportunities to discriminate, not only based on gender, but also based on race, age, and other formally protected categories (Byron 2010; Kalev 2014). In the oil and gas industry, these opportunities are frequent, as companies expand or contract their labor force in response to the ever-fluctuating price of oil. As a result, white women, men and women of color (most of whom are international workers), and older workers face heightened job insecurities. When deciding who to keep and who to eliminate, employers may consider these groups less deserving than young white men of keeping their jobs (Williams 2019).

A number of researchers investigate how the experience of layoffs impacts professionals, including research on their mental health, their sense of self, their families, and their future employment prospects (Gabriel, Gray, and Goregaokar 2010, 2013; Lane 2011; Norris 2016; Pugh 2015; Sharone 2013; Vallas and Cummins 2015). Although respondents expressed of all of these concerns in the interviews, my focus in this chapter is on what their experiences of layoffs can reveal about gender inequality in scientific careers. Through the narratives I collected, we can see how one particular company in the oil and gas industry is organized in ways that are unsupportive if not overtly hostile to women scientists. Oil and gas may be an extreme case of pre-cariousness, but because the key mechanisms I identify are becoming common throughout the new economy, their impact is likely to be far more pervasive than in this single company.

THE LABOR PROCESS IN THE NEW ECONOMY

The increasing precariousness of employment is not the only change in the economy over the past 50 years. The labor process for professionals has also changed considerably. In the past, all jobs or "positions" within a company were organized into a hier-archical and rationalized system or organizational chart. Work-ers who occupied these positions carried out tasks specified in their job descriptions, and they were evaluated and compensated by managers who controlled the flow of assignments. Today, as corporations shed layers of management, work is increasingly organized into interdisciplinary teams. These teams work with considerable discretion on time-bounded projects and are judged on results, often by peers. Membership on teams is fluid, as they assemble, expand, and disband depending on project need. Fur-thermore, in the new economy, standardized career "ladders"— with clearly demarcated rungs that lead to higher-paying and more responsible positions—are being eliminated or replaced by career maps, or "I-deals," which are individualized programs of career development negotiated by workers and their supervisors. These career maps might incorporate an employee's aspirations for a managerial or technical track, as well as family planning, such as anticipated needs for leave without pay or temporary part-time schedules. For job transitions inside the firm, networks have become critical resources. Employees are expected to identify opportunities for their career advancement through their network of personal relationships formed at work and in their professional associations. Networking has also become the principal means through which workers learn of job openings outside their firms (Babcock and Laschever 2003; DiMaggio 2001; Osnowitz 2010; Powell 2001; Rousseau, Ho, and Greenberg 2006; Vallas 2011; Williams, Muller, and Kilanski 2012).

These three elements of the new labor process—teams, career maps, and networking—appear gender-neutral on the surface. Some have argued that because these new elements are more flexible than the older system of standardized career ladders and job descriptions, the new labor process may be more compatible with women's nonlinear career trajectories (e.g., Hewlett 2007). Certainly, customizable career maps seem to enable individuals to craft their own pathways through life instead of following in a lock-step order dictated by the company—a potential benefit for those who want to cultivate interests outside of work. Furthermore, the team structure may resonate with those who prefer collaboration to competition, a quality often associated with femininity in our culture and that many women scientists may find appealing.

However, in this chapter, I will argue that these new features of work also have a downside. During periods of downsizing, each of them injects the possibility of supervisory bias and discrimination into the layoff process. Their informality and flexibility put individual workers at the mercy of supervisors, not all of whom value diversity. At the same time, they shift responsibility onto workers for determining their success or failure. The new labor process absolves the company of any responsibility for looking out after employees' career development and promotes the neoliberal ideal of the free market individual, which has been exposed as a gendered and racialized ideal (Neely 2020).

Under the new labor regime, it is up to individual workers to negotiate, network, and collaborate their way to success. Virtually all of the women in my study felt handicapped in their efforts to do this, but this is especially the case for those who lost their jobs. By delving into their narratives, I illustrate how these seemingly gender-neutral labor processes—teamwork, career maps, and networking—worked to the disadvantage of these scientists.

The first two women whose narratives I discuss experienced "voluntary layoffs," while the second two cases describe women involuntarily laid off from their jobs. The company gave workers the option to volunteer to be severed, which they termed "RFS," or request for severance. This option was not offered with every wave of layoffs, and workers were not guaranteed that their request would be honored. (One of the engineers I interviewed was turned down when he submitted his RFS.) However, if the company granted the request, as they did in the following two cases, the workers received the severance pay they had accumulated (approximately one week of pay per year of service), which they would have forfeited if they had left their jobs without company approval (this was the experience of those discussed in the previous chapter).

The distinction between voluntary and involuntary layoffs clearly mattered to the people I interviewed. Without exception, they considered the option to "self-sever" a humane practice, as it gave workers a chance to preserve their dignity by leaving on their own terms (albeit only with the company's permission). However, I am not convinced that the "choice" between voluntary and involuntary layoffs makes any difference in terms of retaining and promoting women in STEM careers. All four of these women left the oil industry entirely, possibly the reason why GOG paid them to walk out the door.

MARY
"I Just Didn't Feel like I Belonged There"

When I first talked to Mary, she was a 30-year-old mother of two married to another geoscientist working at GOG. She was also one of the company's top-ranked employees, a ranking she shared with her husband. Originally from the eastern United States, she entered college thinking she would major in mechanical engineering or chemistry, but like so many of the others I interviewed, she told me that her love of the outdoors drew her into geology. The great outdoors is a recurrent theme in her narrative, as she and her husband are hiking and camping enthusiasts. They accepted jobs in the oil industry because GOG offered both of them positions in Denver, where they could enjoy plentiful outdoor recreation opportunities. When I first interviewed Mary, she was living in Denver with her husband, and their careers were flourishing.

Avoiding Houston was their top priority, according to Mary, but they knew they would have to move there eventually if they were to enjoy continued career development. Indeed, a year after our first interview, they had moved their family to Houston. Mary's husband received a coveted assignment there, and despite her misgivings, Mary agreed to the transfer. It was a disastrous move for her. After moving to Houston, her interest in staying in the industry plummeted. In 2014, she was 70 percent certain she would stay; a year later she was 80 percent certain she would leave (according to the annual survey she filled out). By the time of our scheduled interview later that year, her certainty about leaving had risen to 100 percent. She and her husband had submitted their RFSs. Moments before we talked, she learned that the company approved her RFS. During our call, she was waiting to hear if her husband's request would also be approved (it was).

What had changed over the course of one year? The short answer is that she and her husband decided to go back to graduate school and pursue their doctoral degrees in geology. The long answer is more complicated, and involves risk, work-family balance, and the quest for a more compatible career.

Explaining her decision to submit an RFS, Mary said, "We were both at the same company, and both tied to the same risky commodity—not a safe position to be in financially for our family." Mary hoped that by earning their PhDs more opportunities for secure employment would open up. Tenured jobs in academia would be their first choice, but if that did not materialize, they would consider jobs in government or consulting—the latter being an option for only one of them. Mary insisted that one person had to have a steady salary and insurance at all times. If that meant she would take a job as a high school teacher while her husband became a consultant, so be it, she told me.

Here we see the allure of female-dominated occupations, a theme that arose in the narratives of other women I interviewed. Nursing and teaching were singled out as alternative careers that would permit women to retain their identity as scientists but that seemed to offer more secure and dependable jobs than working for the oil industry. Mary reasoned that becoming a high school teacher could provide a stable income and insurance for her family, thus providing a critical safety net that could help them manage the insecurity of her husband's career, if he were to work as an oil industry consultant. In this way, the precariousness of industry careers can be linked to occupational segregation.

However, Mary's first choice is to be a professor at a university. She considered academic jobs especially appealing because of their flexible schedules, which seemed much more compatible with raising a family than working for the oil industry. She acknowledged that GOG's approach to work-family balance was inadequate, although she expressed gratitude for the family accommodations the company did make. On the plus side, she praised the optional 9–80 schedule. This arrangement allows employees to work 10-hour days, and take every other Friday off from work. She describes her 9/80 schedule when her children were still in day care in Denver:

> I was able to find one of their preschool teachers who was willing to be a morning nanny for me. So she came into our house around 5:30, we let the girls sleep in, and she would get them up and take them with her to school by 8:30. So we had a really nice schedule, where we could follow the 9/80 and we knew our kids would be sleeping while we were at work in the morning, and then we could pick them up at 4, 4:30.

Mary referred to this "really nice schedule" as an example of GOG's family-friendliness, even though it involved overwork and seeing her children only four hours per day. Like many women professionals in the United States, her expectations of her employer were extremely low, at least compared to her counterparts in Europe (Collins 2019).

Mary recognized that GOG could have done more. Granted, she told me, "We get paid really well. But still, they could be more accommodating." On-site day care would be helpful, for instance. She was also disgruntled about the parental leave policy. GOG allowed her to take only six weeks of paid disability leave after the birth of each child, followed by another six weeks of leave without pay, while her husband was allowed to take one day off. "That to me is asinine," she grumbled. Indeed, a recent study confirms that men benefit from flexible schedules as much as women do (Kelly and Moen 2020).

Mary feared that the industry downturn was making the situation worse for families. Those who negotiated for a part-time schedule after childbirth were at the top of the list for layoffs, she told me, betraying the self-determination promised by well-planned career maps. For everyone else, the contraction of work assignments made it hard to focus and to fill up those ten-hour days with productive work. Comparing her life before and after the downturn, she said:

> [Before the downturn] I really enjoyed my job. I could work hard and play hard with my kids. It was fine. When they went to bed, I could get on the computer and do some work. But when work starts slowing down, or things are not ideal and the environment's weird, then you don't necessarily want to do that and you want to be at home. You are always dealing with that. In terms of management support, I think people have been pretty flexible overall. There are a few managers out there that are by-the-books but some others understand that you have kids: "Please go and work from home." So I appreciate that aspect of what we've dealt with. For every really bad situation I've heard of, I've probably heard of three or four very positive situations.

Mary considered her own situation "positive" on balance, thanks to her "understanding" supervisor. Others were not so lucky, subjected to "bad situations" from "by-the-books" managers. In the end, the inadequacy and the arbitrariness of GOG's family accommodations were factors pushing Mary out of the industry. She was looking forward to becoming a graduate student again, so she could exercise more control over her schedule and spend more time with her family.

A final consideration tipping her decision to "get severed" (as she put it) involved career development opportunities in the industry. When I first interviewed Mary, at the height of the oil boom, she was undecided about pursuing either a technical career track or a career track in management. In response to a survey question, she gave herself a 50–50 chance of taking the management track. By the time of our second interview, however, she was no longer considering that path. If she ever returned to the industry, she told me that she might consider taking a lower-level supervisory position, but she would never consider a top leadership role.

Two experiences helped to convince her that she was unsuited to top management. Ironically, the first one was a leadership-training seminar she attended. The company invited only ten of its "top performers" to this special two-day program. As one of those selected, Mary learned that she was being cultivated for executive positions, "in the VP kind of realm." However, instead of encouraging her ambitions, the event ironically convinced her to leave the industry.

Mary had a hard time putting her finger on exactly what went wrong at the leadership-training seminar. She found parts of the program fun and exciting, and she felt supported by everyone there. However, in the end, she told me, "I just didn't feel like I belonged there." The senior leadership who were present at the event—virtually all "old guys working together forever"—

made her feel like an "outsider coming in." She suspected that she lacked the personality to thrive in this environment: "I'm not as extroverted as most people in that group were, I guess. Every time I take the Meyers Briggs, I'm kind of right in between E [extrovert] and I [introvert]. So there is only so much I can really take before I am overloaded." She admitted to me that she had called in sick and had not attended the seminar on the second day.

Mary attributed her discomfort at the seminar to her personality—an indelible part of who she is that simply does not fit into the world of top managers. This is a textbook example of what sociologists refer to as a person's "habitus," the deep-seated dispositions that reflect a person's early socialization experiences and that play a key role in reproducing social inequality. Mary's gut feeling of being "out of place" is an expression of her habitus. Nothing in her background prepared her to feel at home in spaces with "old guys working together forever." This is how domination works: the "outsiders coming in" blame themselves, and not the company's all-male leadership structure, for their failure to fit in.

A second decisive factor in Mary's rejection of the GOG management track was her observation of women who did "fit in" to that world. Early in her career, she had a manager "who was a very strong woman" and who acted as a role model and mentor for Mary. Unfortunately, this former boss lost her job during the downturn (she was force-retired). In their subsequent conversations, Mary learned just how difficult it had been for this former boss to climb the ranks; she realized that she would be forever stuck in middle management. "It was as far as she was going to go," Mary told me.

In contrast, the few women at GOG who did make it to the top were not good role models for other women. Mary recalled:

There were one or two other women [in leadership] and it seems their whole focus was on their careers. It got a little cutthroat, a little nasty. It didn't feel like once you got to that level, working as a team or bringing people together wasn't necessarily where they were going. It was more pitting people against each other, making it very competitive. Not that I'm not a competitive person, because I am and I know that. But there is a certain limit to what I want to deal with or that I think is healthy.

Mary realized that to make it to the top levels of management at GOG, women had to be ruthlessly competitive, and she never wanted to become like that. Although she wanted to see more women in management, she could not see herself thriving in that environment.

Mary left behind the oil and gas industry and her high salary for the chance to become a professor, a job she considered more compatible with her personality and more flexible for her family. The last time I interviewed her, a year later, she expressed zero regrets about her decision to leave. She told me that her former colleagues at GOG expressed admiration for her and her husband: "They tell me that they wish they had the nerve to do what we did." Although insisting, "I was treated very well at GOG," she was convinced she made the right choice to leave:

The atmosphere before I left was very depressing and almost toxic. I have heard bits and pieces about the last round of layoffs and how difficult it was again. I am glad I was not put through that emotional rollercoaster once again.

Mary's life has been less stressful and has felt more balanced since she left GOG. Her future career is not without risk, but the risks involved in pursuing an academic career seem more manageable, and unlike oil and gas, it is a world where she feels she belongs.

BARB

"The Company as a Whole Has Always
Treated Me Very Well"

I received an email from Barb before the launch of the final wave of the survey in late 2016. She contacted me to tell me that she had become a "statistic," another woman scientist who was leaving the oil and gas industry. A week before a new round of layoffs was announced, employees were given the chance to RFS, and Barb applied. Her request was approved and she contacted me to explain her decision to leave.

When we first talked in 2014, Barb told me that the oil and gas industry was originally a hard sell for her. Growing up in California, she considered petroleum an "evil industry" that "ruins the earth and kills animals." As was the case for so many others I interviewed, a savvy recruiter changed her mind, convincing her that GOG strictly followed voluminous government-imposed regulations to safeguard the environment.

Once employed by the company, Barb found enthusiastic mentors who were encouraging about her career prospects. Although her annual employee ratings were average, she received promotions and raises every year. Wanting to improve her ratings, she designed a clever method to document her accomplishments: She prepared a PowerPoint from her previous performance review, listing her goals for the year, any criticisms she received from management, and then clear evidence that she had addressed each one. She called her method "self-deprecating self-promotion"—an especially innovative approach, I thought, because women often face stigma for touting their accomplishments.

Self-promotion has become necessary for career development as a result of the widespread implementation of the team structure. By the very nature of teamwork, the individual's contribution to the final product is obscured, so to get credit, individuals must be willing to engage in self-promotion. But owing to gender stereotypes, women encounter difficulties when promoting their accomplishments and gaining the credibility of other team members. Many women feel uncomfortable self-promoting because it violates expectations associated with femininity; they walk a line between being perceived as incompetent but likeable, i.e., too feminine, and being perceived as competent but detestable, i.e., a bitch (Williams and Dempsey 2014). Barb's "self-deprecating self-promotion" seemed like a promising approach to navigate this gendered dilemma since it drew attention to her achievements while crediting her superiors for her success. I told her that the technique showed strong potential for management, to which she responded: "That is a very nice thing to say to me. Thank you." A year later I found out that I was the first person to encourage her managerial aspirations.

But my initial interview with Barb was overwhelmingly positive and upbeat. The only potential hitch that she foresaw was her future work-family balance. She wanted children and was not at all sure how parenthood would meld with her demanding career. Much would depend on her supervisor, who would have the authority to approve any special family accommodations. As discussed previously, GOG gives supervisors discretion to negotiate nonstandard schedules with employees as part of their individualized career maps. Barb asked me to turn off the recorder as she detailed her concerns—not that her concerns were unique. On the contrary, virtually all the women scientists I interviewed expressed dismay at the miserly maternity leave benefits at the company (Williams 2017). The only notable aspect of our discussion was Barb's reluctance to criticize the company on the record, even with my assurance of anonymity.

By the time of our second interview 14 months later, this reluctance had disappeared. We spoke while she was on unpaid maternity leave, and she heaped scorn on the company's HR (human relations, or personnel) department. They seemed incapable of providing useful and consistent information for women on maternity leave, she told me. For example, the form they provided for her to fill out to request her leave "is designed for some drunk person who gets hurt on a four-wheeler," not for a new mother. "It was like no woman had ever gotten pregnant and gone out on maternity leave before," she grumbled. The company provided no childcare assistance, either. When she brought this concern up to senior managers, they invited her to prepare a "business case" for why the company should provide child care to its employees—this in addition to her usual job duties. Not surprisingly, she demurred:

> There is a group of us that tried really hard to get on-site child-care.... Our senior management at the time was very responsive to it. Basically they said, which I thought was pretty amusing was/ and these are a bunch of men/ "Ladies we understand your concerns, and we're supportive of it. Why don't one of you draft up a business case for it and we'll review it." And I was like, "Yeah, in all of our copious spare time!" Is that really appropriate for us to be doing? Because you can't/ especially during a downturn, if you look back, you've got someone on your staff who spent six months building a business case for why you should have child care who is a geophysicist—it doesn't make sense. So of course it didn't happen.

Although they did nothing to address the women's concerns, Barb nevertheless described upper management as "very responsive."

Barb blamed the downturn, not the company's leadership, for failing to follow through on their request.

While she was on maternity leave, the company was in the midst of a 25-percent reduction in force, but Barb was feeling relatively secure. They could not lay her off, she reasoned, because she was still on her twelve-week "FMLA-protected" leave, and it would be a "nightmare" for the company if they tried to fire her during this period.[4] However, she was unsure what was going to happen when her FMLA expired. The company announced that anyone who took leave without pay (not counting the 12 weeks of guaranteed FMLA leave) could be fired without severance. Remorsefully, she described a deal she had worked out with her boss when she was pregnant, which would have permitted her to take an additional three months of unpaid leave and then return on a part-time schedule. Unfortunately, the deal was not written down, and that particular supervisor was laid off in a previous round of reductions, causing her a great deal of anxiety. Barb's arrangements would be in the hands of a new supervisor who had yet to be named.

Barb continued to be dissatisfied with her performance reviews. Her innovative "self-deprecating" approach did not work out as she had hoped; she received another average rating despite succeeding on a high-profile project. She expressed confusion over how promotions, raises, and projects were awarded, and frustration at the lack of transparency. In the new economy, as career maps replace standard career ladders, employees have no way to evaluate the fairness of their situations. Because of the individualized nature of career maps, evaluations and professional development opportunities are left up to the discretion of supervisors, who are sometimes motivated by gender bias.

"It was deflating," Barb said, referring to her average rating. "It took all the wind out of my sails." Furthermore, her aspirations for management were squelched when she brought up the topic with her supervisor. The institution of career maps gives employees the impression that they can control their destiny, since they are consulted in making and updating their career development plans. GOG underscores the point by telling workers to "drive your own career." But, Barb said:

> I talked to my supervisor about it last year after speaking to you. I was like, "Why not me as a team lead position? What do I need to do to go about doing that?" And he was sort of taken aback by that. He didn't have an answer for me.

Barb is referring to the same supervisor who was otherwise supportive of her taking an extended maternity leave. She described him as absolutely gushing when he heard the news of her pregnancy; he reportedly said, "That is such wonderful news! I'm so excited for you! I have three kids, it's been a wonderful experience, and I will just enjoy watching you go through this journey." In contrast, he was not at all enthusiastic about her aspirations for management, betraying a gender bias in his plans for Barb.

As a new mother, Barb appreciated all the support she did receive, not only from her boss, but from other people she worked with as well.

> I do feel supported. And there's people who have come out of the woodwork during this tough time of early parenthood. Random people, people without kids, people with grown kids, people with young kids. Coworkers have dropped off food, come by for a visit. One of them came over and cooked us dinner. I've got a gal coming over tonight. She just wants to go for an evening walk with us.

Because I don't know what to do with the baby when he is awake, so I just walk with him.... So I feel well respected in my job and I feel well liked.

Barb experienced plenty of positive reinforcement for motherhood, but virtually none for her geophysics.

Our final interview took place a year later. Her last year at the company was a tumultuous one, as two more rounds of layoffs unfolded. She was very pleased with the new supervisor she was assigned, a woman whom she described as "very supportive." This new supervisor honored the maternity leave arrangement that her previous boss had approved, including a part-time schedule when she returned to work after an extended unpaid leave:

> She said, "Be part-time as long as you feel it's necessary." However, being in a group that drills, I was on 24 hours a day, 7 days a week. So what I was finding was that I was getting paid to work part time, 20–30 hours a week, and I was working 60–80 hours a week. That didn't work for me, so I went back to full time.

Like many professionals, Barb found that her "part-time" schedule did not reduce her responsibilities and resulted in only a marginal reduction in hours for significantly less pay (Webber and Williams 2008). Barb suspected that the company honored her original arrangement only because it saved them money without reducing her workload.

A great deal of shuffling of employees had happened during the time she was away on maternity leave. She lost all of her senior mentors, and she was put on an entirely new team, led by a "golden boy" who had no training in geophysics; he once asked her to "teach me this in 15 minutes" (she refused). Barb suspected that her new coworkers were sabotaging her by undermining her efforts and stealing her ideas. Cycle after cycle of layoffs had produced a hypercompetitive dog-eat-dog environment at work, as everyone jostled to take credit for their team's accomplishments. She started to consider alternative careers.

Two episodes stand out in her narrative explaining her ultimate decision to leave. First, the top manager in charge of geoscientists made a sexist remark to her when she went to talk to him about voluntary severance (he told her she looked ready for a "hot date" when she entered his office). When she complained about it afterward, her immediate supervisor dismissed his comment as "just his personality." Ironically, this top manager, whom Barb referred to disgustedly as a "pig," was the only person who tried to talk her into staying at the company. He told her at their meeting that she was "one of our very best geophysicists and we want to keep you," but it was too little too late: "I was like, honestly, that information would have been useful at any of my midyear or final year reviews in the last eight years." Instead, she said,

> I was told by one boss that I was "needy" and needed to become more intense. I was told by another boss—when I scheduled my own midyear and final year reviews because he couldn't be bothered—that I was doing "fine." Before I went on maternity leave I was told I was doing "fine." There's been no feedback. I've just assumed that I was a middle-of-the-road employee and was never really going to end up doing anything in my career.

Barb saw no future for herself at the company, and described herself as caring less and less about "getting another barrel of oil out of the ground."

The second episode that clinched her decision to leave was a personality test that she took as part of a company-sponsored "team build." Here is how she described it to me:

BARB: So we all filled out this [questionnaire] and found out what colors we are. And then groups of us went into a room and we all discussed our colors and how we could improve on the colors we were deficient in. It was the lamest team build I've ever been a part of.

CW: But on the other hand, you said it gave you insight on what really drives you. ·

BARB: Right. Exactly. It was useful for me because it helped with my decision to leave. I'm of the opinion that you have a personality. You are not necessarily born with it, but you have a personality. And you have strengths and weaknesses. Instead of trying to make those weaknesses your strengths, I think you should just try to be the best at what you are because it's going to come the most naturally to you. Any self-help book that is about how to be more "red" is going to be written by someone who is red.

Personality testing is common throughout the industry, and it is often included as part of diversity training, as it was in this case. At GOG, a mix of "personality types" is seen as adding to the company's diversity. Critics of diversity programs protest that virtually any individual distinguishing feature today counts as "diversity," including one's place of birth, alma mater, and occupation—taking the focus away from addressing the inequalities of race and gender (Collins 2011; Moore and Bell 2011). As indicated previously, the company's "inclusion" initiative institutionalized this broader and virtually meaningless definition of diversity. Without a hint of irony, a mix of "colored" personality types now serves as an indication of diversity at this virtually all-white male-dominated company.

Despite its absurdity ("the lamest team build ... ever"), this test convinced Barb that she lacked the personality to be a geophysicist. "I realized that my personality is not suited to sitting behind a computer and interpreting seismic data," she said. "I want to speak to people, share my ideas, innovate, and make an impact on my community." Stymied from achieving these goals at work, she decided it was time to leave the industry.

Barb is hitting the reset button and going back to school to become a veterinarian. The people she cares about at work reportedly all agree with her decision: "I asked them their opinion, and it was overwhelmingly: yes, you should go do this. It much better suits your personality and it sucks here right now." Fulfilling her desire to control her own destiny, Barb assumed personal responsibility for feeling stymied at work and decided to leave the industry in search of a better psychological fit. Like Mary, she blamed herself, and not her employer, for her failure to fit in, once again betraying the ways that a person's habitus both reflects and reinforces social inequality in organizations.

At the end of the interview, I asked my final question: "Is there anything that I haven't asked that you think is important for understanding your career in the oil and gas industry?" I was taken aback by her answer:

> I am very surprised that the reason that I left was because I decided my skills would be better suited elsewhere. I was always under the impression that I would leave because I was being oppressed as a woman or I was being abused somehow by this industry. But the company as a whole has always treated me very well. Individual people I've had some negative experiences with. But the reason that I left was 100 percent because my skills are better suited elsewhere.

From Barb's perspective, her personality, and not the company, caused her to leave. She has felt supported throughout her career in the oil and gas industry. Yet all of the examples she gave of supportive managers, supervisors, and colleagues were of them extolling her role as a mother—a form of support, it must be said, that is not extended to all women. Black women professionals, in particular, are rarely celebrated for their maternity (Bell and Nkomo 2001). Furthermore, this "support" for motherhood was emotional and relational, not economic, doing nothing to lessen the "motherhood penalty" that women suffer throughout the labor market (Correll, Benard, and Paik 2007).

That Barb was lauded for her maternity and not for her skills as a geophysicist did not constitute sexism in her view. That was the term she reserved for the senior manager who sexualized her and made her feel demeaned. Here we see the allure of individualistic explanations for discrimination, a problem caused by a few "bad apples," not a systemic problem built into the culture or structure of an organization (see also Britton 2017). According to Barb, some people are "pigs," but since personalities cannot be changed, there is no hope for them to change. Likewise, Barb has a personality that cannot be changed. Unfortunately, from her perspective and that of advocates for women in science, it is not suited to a career as a geoscientist in the oil and gas industry.

ALICE
"I Absolutely Should Have Sucked It Up and Not Made It So Obvious How Unhappy I Was"

Alice is a geologist with 10 years of experience in the oil and gas industry. She was laid off from her job in 2016, two months prior to our third interview. Her geologist husband also works for the oil and gas industry, but for a different multinational company, and Alice thought his job was still secure. They are the parents of twins, who were two years old when I last interviewed Alice.

In our first interview in 2014, Alice told me that she became a geologist because she fell in love with the subject in college. A self-described hippie, she never thought she would "sell out" and work for oil and gas. While obtaining her master's degree, she applied to the Peace Corps but was not selected. Lacking other job prospects, she accepted an internship and eventually a position at another multinational oil company. After five years working there, she went to work at GOG to pursue her interest in horizontal drilling.

When times are good in the oil industry, such job changes are common. As is the case for professionals throughout the new economy, geoscientists identify opportunities through personal and professional networks. Without standardized career ladders, employees have few formal ways to identify promotion and advancement opportunities, so they turn to coworkers, mentors, and friends for information and advice about career development. Networks also provide a measure of protection against job insecurity by keying employees into information about jobs in other companies. But networks can be a major source of gender inequality. Typically, women have trouble joining the most powerful, male-dominated networks: In the oil and gas industry, they are not welcome at informal events centered on golfing, hunting, or fishing, for example. Women may turn to one another for support, but in a male-dominated industry, these connections do not reach far up in the organizational hierarchy. Furthermore, women-only support groups may be regarded with suspicion and denounced as frivolous outings, occasions for bitching, or conspiracies against men (Williams et al. 2012). Comparable dynamics impede the career development of racial/ethnic minority men in predominately white organizations (Wingfield 2013).

In Alice's case, virtually all of her job changes—both internal and external—came about because of her personal network of women geoscientists. Her switch to GOG occurred because a woman friend working there told her about the opportunity. Her female supervisor at her old company tried hard to retain her, but was unable to identify a job in just the right area. (Her male manager, on the other hand, said he understood why she wanted to leave and wished her well.) After she moved companies, her first internal transfer came at the suggestion of another female friend, who knew she was dissatisfied with her placement and sent her a job posting that she found on the company intranet.

Alice did not foreground the importance of this female network in her narrative. She did, however, comment on the male domination of her industry. Typically, she worked with only one other woman on teams of twenty or so people. She identified this imbalance as a source of dissatisfaction the first time I interviewed her. She gave herself a 50–50 chance of leaving the industry and the profession altogether, as it was not in line with her desire to "help people." Pregnant at the time with her twins, she thought that teaching or nursing might be a better fit and a more stable and secure job.

A year later, after her twins were born, Alice had a renewed commitment to her career in the oil and gas industry. She was working on a new team with a female supervisor, whom she loved (they were the only two women on their team). Her boss gave Alice permission to work a part-time schedule for almost a full year, which in her case meant working four days a week instead of five. Alice also had the flexibility to leave work early and occasionally to work from home. Out of gratitude, she said she was working longer and harder than ever before:

> At the beginning of my coming back from maternity leave, [the children] were sick like ten different times. One would get sick and

then the other. I would have to miss so much work. My husband could only take off so many days and I could only take off so many days, and we don't have family in town. [My supervisor] has been ridiculously understanding about that.... She lets me leave early if they have a doctor's appointment. And there's no grilling, no questioning, no comments. It makes me want to work harder for her.

In addition to wanting to work harder for her supervisor, impending layoffs at the company added another incentive to perform at a high level. Our second interview took place during the second round of layoffs, and she expressed anxiety about losing her job:

I'm happy where I'm at, I'm enjoying what I'm doing, I have a great situation here. I'm truly anxious about [losing my job]. The first round of layoffs was nerve-wracking, but not as nerve-wracking as this one.... And it's odd, people's behavior becomes very odd.... I cannot tell you how many people, women and men, who have said this to me: "Well, your husband works." OK, so I should just fall on the sword for everybody else because I don't matter? My husband works! He could take care of me! It's a good thing! Yeah, it's been ridiculously annoying.

Alice's experience mirrors that described by Koeber (2002), who found that married women are targets of resentment during layoffs because they supposedly do not need their jobs as much as married men do. Alice briefly considered applying for voluntary severance, but quickly dismissed the idea. Becoming a stay-at-home mother was not appealing to her. She found herself cherishing her time at work as a break from the unending stresses of child care.

Alice still gave herself a 50-percent chance of leaving the company and the industry altogether. Although she loved her job, she was increasingly wary about her future prospects in oil and gas. She felt that her professional opinions were discounted, especially when she dissented from the majority viewpoint. Ironically, she had been told in the past to speak up more in meetings, but when she tried to do so, she was ignored or, worse, penalized for not toeing the company line. In other words, she was caught up in the gendered dilemma of self-promotion. When I asked at the end of that second interview if she had any regrets, she said:

> In times when I wasn't happy with what I was doing, I do wish I had a better attitude. [At the start of my career] I thought, "Well, I'm hot shit here, I can do what I want, they have to want me." And as I see the way the world works, and how things are going, there is nothing wrong with confidence, but I absolutely should have sucked it up and not made it so obvious how unhappy I was. I think in the end, that doesn't help me or any other women because we are just, [in critical voice:] "Oh, she's emotional, she's a woman." And it's unfortunate. So that would be my biggest regret.

Expressing herself turned out to be a liability, not the show of confidence she intended. Instead of gaining her recognition and respect, she felt that making her opinions and desires known undermined her credibility.

The third time I interviewed Alice she was laid off, at home taking care of her twins, and piecing together a narrative to explain what had happened. The first sign of things to come was that her beloved woman supervisor was moved to a different group and demoted out of management. Second, during this major downturn in oil prices, Alice's principal activity—drilling oil wells—was drying up. She kept herself busy with a science project investigating the potential for extracting residual oil from abandoned plays. Although this project was in her career map and had been approved by her new supervisor, it had no deadline, and she had little to show for herself at her end-of-the-year performance review, making her vulnerable to being laid off. And third, she passed up an opportunity to showcase her knowledge to senior management. Her supervisor asked her to organize an internal symposium on horizontal drilling. Upper management would attend, giving Alice the exposure she needed after her year of "unproductive" scientific pursuit. Alice initially accepted the assignment but then balked. She came to see the event as a cynical ploy by managers to help them to decide whom to let go:

> [After I agreed to organize the symposium], I started to think about it: This is horrible! People are going to get laid off in a couple of weeks around here. We know it's around the corner. And I got very uncomfortable asking people to give talks knowing that they could get laid off. Or myself. I got very uncomfortable about setting something up like this and then getting laid off. And so I kind of stewed about it. I got stuff ready. I got it all ready to go. But I didn't send out an invitation. When we heard word that [the layoffs would be announced the following week], I was "Shit! I shouldn't send this now." I self-sabotaged.

Alice said she had experienced an "existential crisis" about whether to send out invitations to the symposium. After planning the event and lining up the presentations, she ultimately decided that she could not go through with it. "I don't have it in me," she said, her voice trembling. "I'm not that way."

From her "20–20 hindsight," Alice believed these three events had precipitated her layoff. At the same time, she was aware that what had happened to her followed a pattern. She noticed that the layoffs targeted older, more "expensive" workers like herself; almost everyone who was retained from her team had fewer than her 10 years of experience. The layoffs were also "heavily skewed towards women having just come back from maternity leave or who had young children":

I'm sure [the company] has people looking at statistics to protect themselves, but there's no way there wasn't a method behind that. There is no way. You are trying to become lean and powerful, so in your mind, no matter what you say, a mom with a small kid is not powerful.

When we last talked, Alice was not sure if she would ever return to the oil and gas industry. She had gone on two oil company job interviews after her layoff—both organized with the help of her network of women friends—but she was turned down for one and she declined a second interview with the other. One thing she knew for certain, she was not suited to be a stay-at-home mother, even though others envisioned that role for her:

> If I hear one more person tell me, "Well, you've got your kids" [crying]. It's not your life choice! It's MY choice.... This is not the choice that I wanted. I was a better mother when I was working. I had so much more patience with my children. This is not happy-home-school-mommy-teaches-them-science-all-day-long. This is preventing catastrophes at every turn.... This is why women had to go through/ what was it in the 1950s? Electroshock therapy? Because they are stuck at home with children. I love them with all my heart, but ... [trails off].

Alice is not certain where her future will lie. When I asked her my final question, she said:

> I don't know. I don't see this getting any better. I see women having to morph.... We'll see what happens. You never know. Maybe I'll find a perfect job, maybe I'll change completely. I don't know.

Unlike Barb, Alice did think people could change. But in the oil and gas industry, only women were required to "morph." Whether she would go back or morph in an entirely different direction was up in the air.

Alice's narrative debunks the corporate shibboleth that having children makes women less productive; in her case, she felt she became more productive after the birth of her children. Regrettably, she may have been penalized for putting her career success above mothering, which violates cultural expectations of women's proper priorities. (Bernard and Correll [2010] label this possibility "normative discrimination.") Alice's narrative also reveals the importance of women's networks to her success in the industry. Other women scientists were the source of her career development in almost every instance. Not that all of the women she encountered were supportive: she claimed that none of the senior women at her company had children, and they were not necessarily sympathetic to those who did. Recall from Barb's account that at this company, maternity was considered a disability, akin to breaking a leg while on a bender.

Alice's supervisor who allowed her to work reduced hours was not a mother, but she did suffer a demotion out of management during the downturn. Although I do not know the circumstances of her demotion, it resonates with Alice's experience of feeling ostracized whenever she expressed a viewpoint that went against the tide. In a white male–dominated company, anyone who assists excluded groups may be cast aside for being a dissenter and a troublemaker, yet another mechanism whereby corporations disadvantage women. When even their allies are pushed out, women do not stand a chance.

CLAIRE
"Why Is Laying Off Women Good for Business?"

Unlike Mary, Barb, and Alice, Claire grew up in an oil family. Both her father and grandfather spent their entire careers working in the industry, her father as a geoscientist for a major multinational corporation. Intending to follow his example, she majored in geology at an "oily" school, one of the handful of universities that act as feeder schools to the giant oil and gas companies. While still in graduate school, Claire went on 20 interviews and received seven job offers, eventually deciding on GOG, which she referred to as her "Goldilocks" choice.

When I talked to Claire in 2014, she was 32 years old and already on the managerial career track. She was working as an assistant to a senior manager of corporate investments. It was a job with tremendous responsibility and exposure: everyone who had held the position before her eventually became a manager. The only drawback of her position, she said, is that it drew her away from the day-to-day practice of geology, a potential liability because her technical skills might diminish while she was in the role. Even so, she was optimistic about her future at the company.

Despite her rapid success, during that first interview Claire expressed a few complaints about GOG. She was unhappy about the lack of women mentors at the firm. She knew no women who were senior to her by 10 to 20 years and who might be able to advise her about her career. She attributed the dearth of women in this age cohort to the industry downturn in the 1990s: "When the bust happened [in the 1990s], no one was hiring any one, let alone women." Although she could turn to her father for advice, she longed to have a senior woman to talk to about the informal gender rules of the corporation, including about seemingly insignificant matters like how to dress and how much makeup to wear at work.

She perceived that the few older women who occupied senior roles at the company had very different attitudes than young women like herself:

They have a completely different worldview when it comes to the industry because they had to claw tooth and nail, and not be women for the majority of their careers. They had to make a lot of sacrifices in family. We [the junior women] come to them and ask them, "Hey would you mind being an advocate because we want to make our work-life balance a little better?" Most of the time their response is "I did it; you need to suck it up and do it yourself."

Claire did not blame these senior women for responding this way. She understood that the male domination of the industry had taken its toll on them. They could not "be women" and survive in the industry. She hoped that younger generations of women geologists would change this; as they amassed power and seniority, they would make the industry more welcoming to women (recalling Patricia's "demographic theory of social change," discussed in chapter 3).

Claire had a two-year-old daughter when I first talked to her, and like Barb and Alice, she expressed dissatisfaction with the scheduling accommodations available to new mothers at the company. Claire mentioned the recent formation of GOG's "parent network," an information and referral listserv. Such employer-sponsored resources are typical of the diversity programs that corporations have implemented to retain women, on the theory that they can provide social support to combat feelings of isolation that may lead to attrition. Implementing a "parent network" requires no resources from the employer, nor does it challenge the company's limited support for new parents. However, by allowing an employee-maintained listserv, the company can bolster its image as family-friendly and welcoming to diversity (Kaiser et al. 2013).

Claire told me that the company explicitly instructed members of the network not to advocate for parents, a perspective she said she understood. GOG "is not running a charity," she said, "it is not up to them to take care of me. They are not responsible for my personal choices." Here she expresses the neoliberal ideal worker norm (Neely 2020), embracing her responsibility to take care of herself and her family (she is the primary breadwinner in her household). Earning over $150,000 per year, she does not need the company's charity. She did think that the company should offer more flexible scheduling options for parents, but she did not think it appropriate to agitate for change, at least not at this point in her career. Change would have to come from another source, perhaps from the young women of today—who are allowed to "be women" and not sacrifice family life—once they become the leaders of tomorrow.

Claire also complained that the industry was hostile to people of color. She lamented the situation of the firm's one African American woman scientist, who was always chosen to represent diversity in the company's recruiting brochures. She is an outstanding geoscientist, Claire told me, yet they insisted on singling her out because of her race and gender, not because of her excellent skills. In her view, the practice of tokenism reinforces the white male domination in the company, a view shared by other critics of diversity programs (Dobbin, Kim, and Kalev 2011; Kanter 1977).

Claire was happy with her job despite these complaints. My impression from that first interview was that her career was on the fast track and the world was her oyster.

Seven months later, Claire was laid off in one of the first rounds of staff reduction. With no foreboding, her manager called her into his office and told her that, due to the downturn, her position was being eliminated. She was immediately escorted out of the building.

When I talked to her in 2015, Claire described the experience of being laid off as "surreal." She had received excellent performance evaluations from her boss, but she found out later that he was not consulted when the decision was made to lay her off. Claire also learned that her position had not been eliminated. A man who had attended graduate school with her was hired to take her place. "Could you have done anything to prevent this?" I asked. The only thing she could think of, she told me, was to change her gender. Women were targeted for layoffs, and she was caught in the purge.

Claire had just begun a job teaching high school math and physics when I interviewed her in 2015. She considered this a temporary arrangement while oil prices remained low. She did not want to be a stay-at-home mother, she told me; she needed the mental stimulation of a job, and teaching was a defensible story to explain her hiatus from the industry:

> So I'm thinking in a year or two from now, all of this will be fairly easy to explain in a job interview. I'm trying to think positively, in case this teaching thing doesn't work out. But I got to tell you, it's a lot of fun. I'm having a very enjoyable time teaching.

Surprisingly, all of her Advanced Placement physics students were girls. Not a single boy qualified for the class. She said that it felt good being a role model for them.

A year later, at the end of 2016, Claire was still teaching high school. I asked her if the passage of time had shed any new light on what had happened to her at GOG. Claire was even more convinced than before that she lost her job due to sexism. Prior to our scheduled interview, she had counted the layoffs among her former associates. Of her 12 closest women friends, 10 had been terminated, a far greater proportion than among the men she knew. She might have been one of the first to go but she certainly was not the last. The reason for her early dismissal, she speculated, probably had to do with her greater "exposure," a term, she reminded me, with both positive and negative connotations. ("Some people die of exposure, you know.") She believed that her heightened visibility as a token woman in a prominent role put her in the company's crosshairs (Kanter 1977). Had she been in a less "exposed" position, she might have lasted another six months, but she would not have survived the purge:

> Eighty percent of the women I knew were laid off. I don't consider myself any different from that eighty percent. Eighty percent of the women who were doing the same thing with their life as I was doing. Early-ish thirties, newish kids. All of those women were laid off. And then more senior women were laid off. I just don't see anything that I could have done differently because I am who I am.

Meanwhile the man who had taken over her job was still in that position. She deadpanned, "He's fully qualified in that he doesn't have a uterus." Counting back 15 years, all of her predecessors in that job remain in their management positions at the company, with only two exceptions: herself and the only other woman who occupied the position previously. She was also laid off.

Claire was discouraged that so many talented women left the company. "Why is laying off women good for business?" she rhetorically asked. Perhaps corporations consider women of childbearing age inherently less productive than men—and they very well might be, she conceded, but only for a short period of their careers. She strongly believed that firing young mothers hurts companies in the end because it makes companies less diverse.

> More diverse teams are more creative. They can see around corners easier. There are lots of dynamics that women bring to the

table, being present in the team structure that men can't provide on their own. But you have to hang in there with women when they are having kids.... It's an investment that some companies devalue. And I think GOG really devalued young working moms in the lay-offs in the past year and a half.

At this juncture, Claire questioned her company's commitment to diversity. The opportunities she had enjoyed early in her career disappeared once times got tough. Her experience convinced her that the company did not value women, despite the rhetoric about diversity.

From Claire's perspective, the company suffers from its lack of gender and racial diversity. Claire was one of the few people I interviewed who both called out and condemned the whiteness of her industry. At the end of the interview, for instance, she told me that she was not worried about her future because she has every advantage that comes from being a white person born in racist America. Unfortunately for her, being a woman got her kicked out of the oil industry.

Claire saw very little hope for the industry changing. Although she knew that some of her older male coworkers were fighting their dismissal by, for example, hiring lawyers to charge the company with age discrimination, she told me that she lacked the time and the energy to complain of gender discrimination.

I just decided to not engage in any of that because to me it was only going to lead to some long battle that would end up in bitterness. It wasn't worth my emotional energy to get caught up in any of that. And I think that's part of the problem. By laying off the young mom, you're laying off someone who doesn't have a lot of spare emotional energy (laughs). You've got these older men who are willing to go to the table and get a lawyer and fight you tooth and nail. But when you lay off a new mom, I'm already stretched thin. So I guess I'll just bow out nicely.

Claire no longer thought that young women could transform the company for the better. They lacked the resources to fight the corporation—possibly another reason why they were targeted for layoffs in the first place, she suggested. She now realized that she was powerless even when she still had a job. She could not protect her network of women friends, nor could they protect her:

> The women, we all stuck together, went out for a beer together, stuff like that. Knew each other's kids. We were there to rally around each other. Maybe that's the reason we got laid off in such high numbers, because we weren't sharing our lives with [the guys]. I don't really know.

Claire's network of female friends, so important for sustaining her hope for a better future, in retrospect might have seemed inappropriate or even threatening to the men at the firm.

Despite her experiences, Claire was undecided about whether she would ever return to oil and gas. She missed the pay, she said, and the "cleanliness" of her old job. Petroleum geology was a career that you could leave at the office, she explained, while teaching enmeshed her in the "messy" lives of her students and their families, which was emotionally overwhelming at times. She has received a number of inquiries from oil companies since her layoff, but she has yet to pursue any of them. If she does return to the industry, it will be at a company that appreciates women. Perhaps she would consider joining a European company, where she hears that conditions are better for women, or else a job in consulting. For now, she is in a holding pattern, content to stay in teaching, where she is underpaid and her talents are not fully utilized, but at least she feels valued.

CONCLUSION

Professionals today face increased precariousness in their careers. Gone is the expectation of a career of 35 years spent loyally working for a single employer. Other features of professional work have also changed. The work process is now typically organized into teams; individualized career maps have replaced standard career ladders; and networking has become the primary means through which future opportunities are identified. These innovations reflect the new organizational logic that characterizes work in the twenty-first century (Acker 1990; DiMaggio 2001; Kalleberg 2018; Williams et al. 2012).

Although they are gender-neutral on their face, these features of work can stymie women's professional career development and contribute to gender inequality in the new economy. Drawing on the unique experiences of four women geoscientists who left the oil and gas industry, the gendered consequences of teamwork, career maps, and networking are revealed. First, the self-promotion required in the team setting puts women at a disadvantage. They are told to speak up and make their opinions known, but because of gender stereotypes, they may be punished if they do so for being too demanding, forceful, or emotional. Second, the institution of career maps gives the impression of enhanced autonomy in professional jobs, but these individualized plans obscure the criteria used in evaluations and provide ample leeway for supervisors to discriminate against women. The lack of transparency in their evaluations, promotions, and family accommodations caused these women a great deal of stress; they perceived that positive outcomes depended almost entirely on having a supportive supervisor, or at least one who does not harbor gender biases. And third, to the extent that it is segregated by gender, networking does not necessarily enhance women's careers. All four women spoke of their reliance on networks of women for advice and support, but these networks could not protect them from layoffs, and in the end, those who supported them met similar fates.

In addition to furthering our understanding of the mechanisms that reproduce gender inequality in the new economy, their narratives cast renewed doubt on the depth of corporate support for diversity (Collins 2011; Williams, Kilanski, and Muller 2014). GOG's endorsement of diversity was a source of optimism for these women early in their careers. But over time, as they faced the possibility of losing their jobs, they began to question that commitment. The traumatic experience of layoffs upset the worldview that had previously bolstered the legitimacy of the corporation in their eyes. The women's taken-for-granted expectations—that their success would be recognized, that the company cared about diversity, and that they drove their own careers—were upended. Granted, in two cases, the women scientists left "voluntarily." Mary thought she was a better fit for academia, while Barb, who was not "being oppressed as a woman ... or abused somehow by this industry," thought her talents were better suited to veterinary medicine. Mary received recognition for her scientific talents, but she felt like an outsider in the upper rungs of the company, and blamed herself for not being able to take advantage of the opportunities offered to her. Likewise, Barb told me that she was surrounded by "supportive" bosses. However, as her narrative details, these bosses lauded her for motherhood, while making no accommodations for her child's care. No one seemed to respect her scientific and management skills, except for one "sexist" boss, and he told her so only after she had made the decision to leave.

From Mary's and Barb's perspectives, "personality" is the reason why some are destined for successful careers in the oil and gas industry and others are unsuited and must leave, an explanation fostered by the individualizing culture of the firm (and by diversity programs in general). Their beliefs in unchanging character traits were confirmed by corporate-sponsored personality tests, which ultimately convinced them to pursue different careers. In this way, they were encouraged to take full responsibility for abandoning the industry, despite all the impediments the corporation imposed on them.

Mary and Barb volunteered to leave but they did not "opt out" (Stone 2007). Nor did Alice or Claire. Looking at these young mothers as statistics, it may appear that they left their scientific careers "for the family" (Damaske 2011), but nothing could be further from the truth. According to Alice and Claire, they worked harder once they had children; their jobs became more important to them, not less. Moreover, they described themselves as scientists, devoted to geology and temperamentally unsuited to the stressful and "messy" demands of stay-at-home motherhood.

Although Mary and Barb did not blame gender discrimination or oppression for their decisions to leave, both Alice and Claire said they lost their jobs due to sexism. Alice knew she was on the chopping block because she lacked exposure at the company. When she finally got the chance to step into the limelight, she faced an existential crisis and decided she could not do it—a decision that makes sense given her past experience at the company, which had penalized her for standing up for what she believed. As a result of her lack of exposure, she lost her job. Claire, on the other hand, had too much exposure. She was convinced that the company was purging women, and her heightened visibility as a token woman on the management track made her an easy target for early dismissal.

The accounts I gathered reveal the mechanisms that stymie the careers of women scientists in the oil and gas industry. In a male-dominated environment, teamwork, career maps, and networks institutionalize bias against women even when companies claim to support diversity. Adding women when times are good does nothing to alter these organizational processes. When it came time to choose who to retain and who to let go, these young women scientists did not stand a chance. As Alice remarked, the company was "trying to become lean and powerful ... [but] no matter what you say, a mom with a small kid is not powerful." And in a company that censures dissent, neither are their advocates.

The women I interviewed did not perceive the few senior women leaders at the company as advocates. On the contrary, these women were characterized as cutthroat competitors who were not allowed to "be women." Senior women at the company did not have children, nor did they support those who did, according to my respondents. The four women in this chapter resolved to never become like them. Observing the women in leadership helped to convince Mary to leave the industry, while prompting Claire to work hard for a chance to replace them.

It is deeply disappointing that none of these women found women allies at the top of the firm. On the other hand, it is not surprising that a company that is overtly hostile to women would select women for leadership posts who are themselves hostile to women. Being ruthless toward other women can be a loyalty test for senior women to prove themselves worthy of leadership positions. Meanwhile, those who advocate for women fall by the wayside.

The oil and gas industry is an extreme case, characterized by high degrees of both instability and male domination. However, as instability becomes the new norm for professional careers, it may be the harbinger for the rest of the economy. The narratives of Mary, Barb, Alice, and Claire reveal that, even in the era of diversity programs, corporations are gendered organizations that discriminate against women. To ensure that precariousness does not further entrench gender inequality, the key mechanisms of teamwork, career maps, and networking must be monitored and scrutinized for their role in reproducing male domination in the new economy.

CHAPTER SIX

Organizational Gaslighting

In 2020, the US oil industry was in trouble again. In the midst of the coronavirus pandemic, the price of oil fell into negative territory, meaning that for a brief period, it cost more to store oil and gas than to take it out of the ground. The situation became so dire that the political allies of Texas oil companies contemplated imposing limits on oil production, a heresy to the fundamentalist free market orthodoxy that rules the industry. The proposal was defeated 2–1 by the state's oil and gas regulatory body, but commissioners did unanimously agree to roll back Texas's already lax environmental regulations on the industry. Meanwhile, oil industry executives oversaw a new round of layoffs.[1]

Throughout its history, the oil industry has responded to fluctuations in the price of fossil fuels by laying off its workforce. This routine business practice has been unquestioned for decades, although it results in periodic hand wringing by corporate leaders about the dreaded "crew change." However, a few in the industry are beginning to wonder if this practice is sustainable. They are questioning the industry's treatment of STEM workers as "disposable labor," as in this much-shared LinkedIn post by an industry consultant with a PhD in geophysics:

> I occasionally give geophysical talks to students in nearby universities, and I have to admit that this is the first year where I recommended that they not consider this industry as a career option. I realize that every industry goes through changes and we all need to adjust accordingly. However, this industry has a very nasty record of treating high-end professionals as disposable labor. There is simply no logical argument to advise young scientists and engineers to invest 5 to 8 years in school to enter a field where they will get laid off every two to three years, and during the time they are employed, suffer constant anxiety of waiting to be laid off. (McColgan 2019)

Subjecting its technical labor force to constant anxiety mirrors the industry's overall approach to doing business. The oil industry is notorious for externalizing the costs for the damage it wreaks on the environment and surrounding communities. Its personnel policies reflect the same approach, as companies sacrifice the well-being of workers in the interest of short-term profits. Not only is destruction wrought on the outside; those on the inside bear the emotional costs of the industry's approach to externalizing risks.

This book has focused on the experiences of women scientists in the oil and gas industry. In US society today, women scientists are having a moment in the sun. Women and girls are encouraged to enter STEM professions in order to make up for a much-hyped labor shortage, which may not even exist (Teitelbaum 2014). Publicity campaigns for women in STEM highlight the exciting and socially important work they do. Women scientists are glamorized in popular culture, where they are depicted as engaging in cutting-edge, necessary, and socially beneficial work, and passionately committed to solving the world's most pressing problems.[2]

Scholars contribute to this glamorization of women scientists by emphasizing their passionate commitment to their work and the critical role they play in undermining sexist stereotypes. Importantly, these studies mostly focus on women scientists working in academic contexts. In the university context, it is perhaps easier to frame scientists' contributions in positive terms. But most women scientists (like most men scientists) do not work in the academy. They work for private employers. By sanctifying women in STEM, we may forget that they may be designing weapons systems, concocting addictive drugs, developing unhealthy foods, and extracting fossil fuels.

Many of the men and women I talked to genuinely believed that they were contributing to the social good. Patricia, one of the "stayers" profiled in chapter 3, said this in response to my question about what she likes most about working in the oil and gas industry:

> I like the opportunity to be able to create energy or bring energy to the market safely. So when I was in college, I was a tree-hugging hippie. Oil companies were going to be bad. But then, the more I thought about it, I could do more good on the inside than the outside. So the idea that every day I get to go to work and we're in the business of creating energy and bringing energy to market, and we are going to do it safely—makes me very happy.

Her response comes right out of the oil industry's playbook. It may be a controversial position to take in the current geopolitical context of global warming, but Patricia sincerely believed it. She willingly abandoned her "tree-hugging hippie" roots to work for GOG, a company that poisons the environment, concentrating its worst damage in poor and minority communities in the United States and abroad. GOG also lobbies for drilling access to pristine wilderness areas and for the right to pump oil and gas over indigenous lands. Patricia is well aware of these issues—she is, after all, a geoscientist with expertise in hydraulic fracturing—but she decided that she "could do more good on the inside than the outside." Her career is spent designing "best practices" for safely fracking the earth.

Many feminists share Patricia's hopeful view. Advocates for women in STEM maintain that hiring more women will make toxic industries become more socially and environmentally responsible. This is the argument made recently in a report issued by the Committee on Sustainable Energy of the United Nations Economic Commission for Europe:

> The business case for greater gender diversity in the extractives industry is compelling. A growing body of research shows that greater female inclusion provides a larger pool to meet the high demand for laborers, higher retention of key talent, increased profitability, better performance, improved safety records, higher standards of government behavior, and benefits to women and the broader community. (UNECE 2019:9)

The report grandiosely claims that most of the problems caused by the oil and gas industry could be solved if more women were calling the shots on the inside (see also Perks and Schulz 2020).

Viewing women as the natural stewards and saviors of the environment also underlies corporate "greenwashing" publicity campaigns (Bell et al. 2019). Oil companies routinely use women in their advertisements to signal their commitment to environmentalism and social responsibility. These ad campaigns draw on popular stereotypes about women and femininity, but do not promote women's interests. Instead, they are cynical ploys to escape culpability for the systemic inequalities they produce by making it appear that they support women (see chapter 1).

Don't get me wrong. I believe that there should be more women employed in the oil and gas industry. The underrepresentation of white women and the almost complete absence of men and women of color are evidence of strong social closure mechanisms in operation. Oil companies are currently organized in ways that concentrate power and resources in the hands of elite men (mostly white men in the United States). However, adding women into an existing structure designed by and for these elites will not transform the industry. Instead, it is far more likely that this powerful industry will transform all who enter it.[3]

The young scientists entering the company I call Global Oil and Gas are paid handsomely to find fossil fuels lodged deeply in the ground and under the oceans. The company spends a great deal of time, money, and effort convincing them that the oil industry is an excellent employer and a stalwart protector of the environment (see chapter 2). These arguments find an eager audience among graduate students straining under mountains of student loan debt. In addition to high salaries, the company promises these young scientists access to vast amounts of data and cutting-edge technology, appealing to their professional identities and quelling their concerns about working for an "evil" industry. Upon joining the company, they are assigned to interdisciplinary teams with other young people and asked to collaborate to solve vexing technical problems. They are given responsibility to find and drill for oil and gas. They are encouraged to drive their own careers and to form personal and professional networks to identify future opportunities inside the company.

This inviting, almost idyllic image of the scientist-at-work draws many into the industry. Unfortunately, it can also inure them to the harms their work can inflict on society and the environment. This is not by chance. Anthropologist Stuart Kirsch (2010, 2014) argues that toxic industries have an interest in monopolizing scientific expertise; offering lucrative paychecks and desirable working conditions are strategies to lock in the best talent. The oil industry also uses its vast wealth to pay university researchers and government lobbyists to portray the industry in a favorable light; oil industry–funded research that debunks the association between fossil fuels and global warming is one example. Moreover, Kirsch notes that it is now standard corporate practice in the industry to impose confidentiality agreements on employees and to limit the access of independent scholars to reviewing its data and research findings. By manipulating science and scientists in these ways, corporations successfully manage their critics, mislead the public, and avoid government regulation.

In addition to the strategies Kirsch uncovers, oil companies use periodic downsizing to control scientists and their work. In my study, I found that the constant threat of layoffs can breed abject conformity, as the fear of losing a coveted job keeps workers quiescent, unwilling or unable to object to dubious corporate practices (see also Beamish 2002; Kelly and Moen 2020). In any industry, churning the labor force for short-term profit is a deplorable and destructive employer practice. That this is occurring in one of the most toxic industries on Earth makes it particularly dangerous, however. In the oil and gas industry, silencing the geoscientists is potentially catastrophic, as these workers are uniquely qualified to understand the poisonous consequences of fossil fuel extraction. If the company makes them unwilling to "safe share," as Ben indicated was the case (see chapter 3), the entire planet is at risk.

Scientists working for the oil industry have an additional disincentive to speak out in the midst of a weak labor market, as they fear gaining a professional reputation as a troublemaker. Even those who were laid off told me that they did not want to burn any bridges leading back into the industry. Despite dire warnings of a shortage of STEM talent, opportunities for geoscientists contract when the price of oil falls, which it seemingly inevitably does. When this happens, older workers are in danger of being replaced by cheaper, younger scientists fresh out of school. International workers are vulnerable to losing their visa status and being sent "home" to their countries of origin, where professional opportunities may be scarce. For all of these workers, their personal stakes in conformity and keeping quiet are intensified by layoffs.

Layoffs are also a factor in understanding the continued white male domination of the oil and gas industry. This book has focused on the careers of mostly white women scientists who came into the industry during a boom in oil prices amid global apprehension about "peak oil." Their racial privilege got them in, but their gender was a factor in kicking them out. When the industry needed women, which it did for a brief period during the diversity craze and oil boom of the 2000s (Cai, Maguire, and Winters 2019), companies instituted programs to recruit and retain these women. But when the industry contracted, they were among the first to go.

The routine use of downsizing gives managers carte blanche to discriminate (Byron 2010). Because there are virtually no federal restrictions on layoffs, managers can make their decisions based on their biased assessments of who is most deserving to keep their jobs. They may point to "merit" and "skill set" to justify their choices, as they did at GOG, but these seemingly objective metrics can simply cover up managers' gendered and racialized preferences (cf. Kalev 2014). They offer excuses but not evidence to justify eliminating people's jobs. Lacking any kind of oversight, managers can lay off women and their advocates with impunity. To protect themselves from public scrutiny and lawsuits, companies withhold severance pay from laid-off workers unless they promise not to sue the company for discrimination.

Economic downturns can be a litmus test of the genuineness of a commitment to equality and diversity. This was the conclusion of British sociologists Maria Karamessini and Jill Rubery (2014), who studied the immediate aftermath of the 2008 economic crisis in Europe. They found that in some countries, "gender issues [were] allowed to slip back or even off the agenda altogether," revealing that for these governments, equity policies "are at best a distraction" (334). Especially in those national economies based on the male-breadwinner model, they conclude that commitments to gender equality were only "weakly embedded" in public policy (334).

Although Karamessini and Rubery focus on gender in international politics, their conclusions also apply to inequality in multinational corporations. When the downturn began, GOG revealed that "diversity" was a weakly embedded corporate goal. In the midst of several rounds of layoffs, GOG cynically replaced its short-lived efforts to promote diversity with a new emphasis on "inclusion" (see chapter 3). As the new approach was explained to me, the inclusion initiative sought to increase "diversity of thinking" and so-called "cognitive diversity"—unmeasurable and ineffable qualities that do nothing to challenge the white male domination of the industry. That this new approach was enthusiastically promoted by Lynne—one of only three women scientists in my sample who remained at the company for the duration of the study—says a great deal about how companies use layoffs to manufacture consent and stifle criticism among employees, as well as the effectiveness of organizational gaslighting.

Acknowledging the dearth of women in professional positions, some of the women I interviewed pointed to individual sexist bosses (both men and women) who stymied their career development. Individualistic explanations like this lead to individualistic solutions, such as ineffective "unconscious bias" training to root out managerial and coworker prejudices. Others embrace what I call the "demographic theory of social change," an unfounded belief that less-sexist and less-racist young people will eventually replace those in the powerful "good ole boy network" whose old fashioned attitudes block progress for women and minority men.

Scholarly efforts to explain the overrepresentation of white men in the industry sometimes emphasize individual choice as well. For example, some scholars attribute the absence of women and minority men from the industry to their personal preferences: these groups simply choose different majors in college, have no interest in working for the oil industry, lack "passion" for leadership roles, or leave the industry to pursue other careers or interests that are a better fit for their "personality types."

These individualistic explanations ignore the institutionalized discourses and practices that produce and reinforce social inequality in the oil and gas industry that are documented in this book. They do not address the ways that the discipline of geology appeals to the interests and the experiences of white students. They do not challenge the recruiting practices of oil companies that target their efforts at predominately white southern institutions, where they lavishly fund internships, fellowships, and professorships. They do not identify the sexism built into performance reviews that punish those with parenting responsibilities. They ignore the ways that teamwork, career maps, and networking systematically disadvantage women. And they do not acknowledge that routine layoffs enable companies to wipe out any diversity they achieve during boom times, while breeding abject conformity among those who remain.

Steven, whom I profiled in chapter 4, perceived that he and the other young scientists in his cohort were misled at the start of their careers by duplicitous recruiters. They were told that the industry's leaders had learned their lessons from previous downturns, so their jobs would be stable, and because of the "crew change," they would enjoy unlimited opportunities for advancement. Similarly, the women I interviewed genuinely believed that the company valued them and their contributions. They were gaslighted by diversity discourse. But when the downturn happened, their early optimism turned into despair. "Why is laying off women good for business?" asked Claire, an early victim of downsizing (see chapter 5). After witnessing ten of her twelve women colleagues lose their jobs, she struggled to understand why the company was so hostile to women.

For the most part, the men I interviewed did not perceive that women were being targeted for layoffs (although a few speculated that mothers were choosing to leave). The men and women did agree that older workers were being let go, but because the rubric of "retirement" was used (for dismissing everyone over 50!), they did not find the practice especially problematic. Few mentioned the impact of downsizing on other groups. One exception, a white male engineer, said this when I asked him how the layoffs were impacting diversity at GOG:

One day a woman who used a wheelchair, a Black man from Africa, and one of our gay colleagues were all laid off on the same day. It's like that was the hat trick of diversity layoffs in those three people. I don't feel like they were targeted necessarily, but it just so happened that one day.... It certainly doesn't give good optics to those of us who pay attention to those kinds of things.

He was an exception, however. At GOG, he said, only outcasts care about diversity. In fact, his colleagues label him a "socialist" (and not in a good way) for voicing his unpopular opinions on the matter.

In my interviews, I never encountered anyone who told me that they were opposed to diversity. Expressing such views to might have been considered offensive, since I am a sociology professor at the University of Texas at Austin, considered by many Texans to be a bastion of liberalism. However, diversity was a topic of heated debate on the website *TheLayoff.com*, a wiki that enables employees (or people posing as employees) to anonymously post comments and rumors about layoffs at their company (the website is searchable by company name). I followed postings about GOG on this website over two years, and observed vicious hostility expressed toward women and international workers—the two groups that represent "diversity" inside the company. Some posts blamed the industry downturn on misdirected efforts to increase diversity, seen to undermine the company's stated goal of hiring the most talented and most qualified employees (Williams, Kilanski, and Muller 2014). During the downturn, these postings were viewed upward of a thousand times.

Thankfully, anonymous websites are not an accurate gauge of public opinion. However, the vitriolic comments posted on *TheLayoff.com* attest to the circulation of these repugnant opinions throughout the company. Expressing them openly may be considered impolitic and possibly taboo, but that does not mean that they are unconscious or rare. Instead, these opinions fester in the anonymity of the chat room, where they condone and naturalize the practices that reinforce the white male domination of the company.

White women's underrepresentation in the oil and gas industry and the near absence of Black, Latinx, and Native American men and women in professional positions are the result of corporate practices. Companies may claim to support diversity and inclusion, but their efforts in this regard are best understood as forms of organizational gaslighting. The policies they implement do nothing to disrupt systemic sexism and racism. Through routine personnel decisions involving recruitment, promotions, and layoffs, oil and gas companies systematically perpetuate the white male domination of the industry.

Why do they do it? Why do companies that perpetuate these discriminatory attitudes and practices persist? According to economists, discrimination is irrational because it is not profitable. Economists assume that employers who indulge their preferences for white men have to pay a premium for their labor, cutting into their profits if they lose meritorious (and often cheaper) workers to their competitors. From this perspective, it makes no sense that GOG would go through enormous effort to recruit women, only to chuck them out when the price of oil falls.

It makes no sense, that is, unless decision-makers at GOG perceive women scientists to be less capable, less productive, or less deserving of keeping their jobs than men scientists. This is in fact a common stereotype.

TABLE I

The Binaries of Science

Positive	Negative
Systematic	Unsystematic
Abstract	Contextual
Universal	Particular
Objective	Subjective
Neutral	Biased
Rational	Emotional
Logical	Intuitive
Hard	Soft
Modern	Primitive
Enlightened	Backward

Twenty years ago, social theorist Patricia Hill Collins (1999) argued that science is a deeply gendered and racialized enterprise. Because white men dominate the scientific establishment, their interests are embedded in science in a variety of ways: in the choice of problems to investigate (the "what" of science); in the design and interpretation of experiments (the "how" of science); and in the definition of science itself, which is the focus of her analysis. To make this argument, she cataloged the binaries that underpin the scientific worldview (elaborated in table 1). The characteristics listed on the left hand side of the table are those associated with the scientific perspective, while those on the right describe its binary opposite. Thus, to be scientific means to be abstract, rational, and objective; to be nonscientific is to be guided by context, emotion, and subjectivity. Many scientists would look at this list and agree that it is an apt definition of science (although some might admit it is an ideal that is often compromised in practice).

TABLE 2

The Binaries of Hegemonic Masculinity

Positive	Negative
Systematic	Unsystematic
Abstract	Contextual
Universal	Particular
Objective	Subjective
Neutral	Biased
Rational	Emotional
Logical	Intuitive
Hard	Soft
Modern	Primitive
Enlightened	Backward

For Collins, however, this definition of science is best understood as an ideology that reflects and reproduces the power of elite white men. How so? The list of characteristics associated with "science" also describes popular stereotypes about elite white men. All are associated with hegemonic masculinity, characteristics often ascribed to elite men (see table 2). The list comprises a historically specific set of beliefs and practices that support and legitimize white male domination. Feminized and racialized groups have difficulty occupying this standpoint, as we are the "others" against which hegemonic masculinity is defined. Instead, we are stereotyped as possessing the same characteristics as those listed in the "not science" column. Collins's analysis helps to explain why women and men of color are underrepresented in science, being perceived as incapable of personifying objectivity, neutrality, and abstraction. Understanding this dominant definition of science as an ideology of gender and race thus helps to explain how it reinforces the white male domination of the scientific enterprise.

This definition of science also provides scientists (and their employers) with a rationale to exploit natural resources and people around the world. In addition to overlapping with stereotypes of gender and race, this conception of the "modern scientist" functions as the binary opposite of the "unenlightened primitive," and supports the spread of industrial capitalism around the globe. According to this logic, scientists and engineers bring enlightenment and progress to "backward" and undeveloped places, including mostly poor and minority communities in the United States and resource-rich areas of the global south. In other words, the discourse of science promotes the interests of multinational corporations in exploiting resources and people around the world.

Collins's analysis helps to explain why companies that spend so much effort recruiting women would be quick to lay them off when the price of oil falls. Managers draw on these stereotypes whenever hiring, promotion, and firing decisions are made in this industry. When jobs are plentiful, companies are willing to hire white women. We saw this at GOG, as the cohort of new scientists and engineers in this study reflected the gender composition of new graduates. The industry may value the ability of white women to "purify" a sullied industry, but the growing percentage of white women graduates coupled with a hiring boom ultimately forced the company's hand. Very few racial/ethnic minority women or men were hired, except for those from foreign countries where GOG had drilling interests. The industry focuses its recruiting efforts at predominately white institutions, guaranteeing them a predominately white workforce (see chapter 2).

However, when jobs became scarce, and managers were told to cut employees, they were empowered to act on the stereotypes that Patricia Hill Collins identified. They could define women of all backgrounds, as well as racial/ethnic minority men, as less capable of science than white men. Even though their performance ratings were virtually identical to men's, these groups were defined as less deserving of keeping their jobs (Williams 2019).

This is not the only study that found that layoffs are an opportunity to discriminate. Sociologist Reginald Byron (2010) concluded in his study of 11,000 discrimination cases in Ohio that pregnant women and Black men are the two groups most often targeted for termination. His research shows that when employers have no accountability—which is often the case in "right to work" states—gender and race stereotypes inevitably will shape their decisions about terminating employees (see also Kalev 2014, 2020).

This study has focused on how elite interests are served inside companies like GOG, which privileges white men as its ideal workers. However, the industry also drives social inequality outside the walls of the workplace. Neither the profits nor the externalized costs of oil and gas extraction are shared equally by members of society. The massive wealth from oil and gas benefits elite investors, while pollution generated by this industry does most damage to the health and well-being of minority communities around the world, with a disproportionate burden of these harms borne by women (Bell, Daggett, and Labuski 2020; Daggett 2018).

The oil and gas industry is only one sector of our economy, albeit one that has outsized political, economic, and environmental importance. It has experienced extreme cycles of boom and bust throughout its 100-year existence. As an extreme case, it shows what can happen when downsizing becomes a routine and legitimate corporate practice. Any society that permits wanton layoffs—whether in response to economic crises or to promote shareholder interests—will guarantee the concentration of corporate control in the hands of a white male elite. Layoffs also produce a labor force cowed into submission by fear of losing their employment. Preventing scientists from voicing their opinions is a recipe for global disaster. A diverse technical workforce empowered to dissent without fear of arbitrary layoffs is the best hope for an accountable oil and gas industry.

METHODOLOGICAL
APPENDIX

This study had its genesis in a random email I received in 2008. I was contacted by a member of PROWESS—Professional Women in Earth Sciences—to help them analyze survey data they collected to figure out why women geoscientists were leaving the oil and gas industry. PROWESS, a standing committee of the American Association of Petroleum Geologists (AAPG), consisted of two dozen prominent geoscientists, most of them white women, working in industry, academia, and government. Earlier that year, members of PROWESS had designed and administered an online survey of geoscientists to gather data on women's attrition from the oil and gas industry. They knew from their personal experience that women were leaving the industry. As one member of PROWESS told me, "We know women have left the industry. The challenge of the survey is to change things." Perhaps because of their scientific backgrounds, they were confident that by gathering data to document women's attrition, companies would be forced to address the problem and to alter their practices to prevent women from leaving the industry.

In retrospect, it was naïve to believe that companies cared about losing women. It was also naïve to assume that women were choosing to leave their lucrative STEM jobs. In the volatile oil and gas industry, the decision to stay or leave is not always theirs to make. Oil and gas

companies routinely lay off workers when the prices of these commodities fall—a practice that would become all too obvious a few
years later. Going into this study, however, my focus as well as that of
the PROWESS committee was on understanding why women scientists were deciding to leave.

AAPG's Corporate Advisory Board (made up of industry representatives) supported the idea for the PROWESS survey, but the
money they pledged for a survey design consultant never materialized, which is why members of PROWESS took it upon themselves to
write the survey. Exxon/Mobil, however, did agree to sponsor a luncheon at the AAPG annual convention to publicize the results—a mere
six months after the survey launched. That is why PROWESS urgently
requested my help in interpreting their survey results, and why I in
turn contacted my colleague Chandra Muller, who is an expert in survey analysis (I am not). We jumped at the chance to study a powerful
industry that is typically closed to outside researchers.

The PROWESS survey generated over 2,000 responses from all
over the world. However, the survey had fatal flaws. It omitted basic
information, including the respondents' gender, age, race/ethnicity,
citizenship, family status, occupational status, and education. Moreover, we had no way of knowing who did and who did not answer
the survey. PROWESS distributed it to all 35,000 AAPG members
with instructions to forward it to women in their personal and professional networks, so respondents were not limited to AAPG members.
Open-ended comments indicate that several men answered the
survey.

Interpreting the survey results was even more challenging for us
because we were newcomers to the industry and we didn't know the
meaning of some of the terms they used in the survey. For instance,
the survey asked, "Why do you think most women leave the energy
industry?" We didn't know what "the energy industry" referred to
(they meant "oil and gas companies"). Thus, in order to interpret the
results of the survey, we first had to clarify the basic assumptions
behind the survey questions. To do so, I interviewed each member of
PROWESS over the phone in April 2009. Only then could Chandra
and I, working with graduate student Jessica Dunning-Lozano, iden

tify themes from the survey to present at the AAPG conference two months later.[1]

Frustrated at the limitations of these data, Chandra and I decided to launch our own study of women's attrition from the oil industry. We worked together on a grant proposal to the National Science Foundation (NSF). To prepare for that, in 2010 we conducted 30 in-person in-depth interviews with women geoscientists with at least five years' experience working in the oil and gas industry. We found respondents with the assistance of the PROWESS committee, and from the AAPG presentation, where we gathered business cards from women willing to be interviewed. The 30 women we interviewed worked for fourteen different firms, including all of the majors. As a result of that round of interviews, we published two papers with graduate student Kristine Kilanski, in which we laid out an approach to understanding discrimination and diversity in the oil industry, and submitted a full proposal to the NSF (which was ultimately unsuccessful).[2]

These interviews revealed the strong impression that women left the industry in their first five years, a finding confirmed by research on women in STEM published a few years later (Glass et al. 2013). Because women seemed to be most vulnerable to leaving at the early stages of their careers, we designed a study of newly hired scientists in which we would check in on them annually over the course of five years. This longitudinal design, we hoped, would capture their decision-making as it unfolded. By following them over time, we thought we could identify the different factors that led some women and men to decide to leave and others to stay.

We approached several different companies to participate in our study, visiting their corporate headquarters in Houston and elsewhere. Ultimately, a member of PROWESS arranged for our access at GOG, where a personnel manager supplied an email list of all of the approximately 360 scientists and engineers hired by the company in the previous five years. (We included engineers in the study as a comparison group.) Managers and supervisors at the company further instructed us on the technical jargon used at the firm, which enabled us to write customized survey questions for the employees of GOG. The survey itself was written and administered without corporate oversight.

Of the approximately 360 people in the cohort we studied, about 70 percent were men and 30 percent were women. The overwhelming majority were white US citizens. In this cohort, 63 people (about 18 percent) described themselves as Black, Hispanic, or Asian/Pacific Islander. However, only 20 of the 63 people in this group were US citizens, and most were engineers. The entire cohort of 360 people includes only two Latinx geoscientists, but not a single African American geoscientist—an extreme instance of what sociologists call "social closure" (Weeden 2002).

At time 3 (2014) of the survey, we asked respondents to volunteer for a follow-up interview, and 57 of them did so; ultimately I interviewed 44 of them (the others did not respond to repeated requests). These respondents included five individuals (three men and two women) who had left GOG since the survey began.

This interview sample of 44 volunteers included 19 women and 25 men. Their average age at the beginning of the survey was 33. All but seven were white US citizens. Of these seven, five were foreign-born (three Latin Americans, one Asian, and one African), one person identified as Native American, and one as Hispanic.

I worried that only people who had positive experiences at the firm would agree to talk to me, a pitfall of all studies that rely on volunteer respondents. I sensed a reluctance, even a taboo, among several respondents about saying anything negative about the company. I was asked on a few occasions to turn off the recording when the respondent brought up a complaint about the firm. One white male engineer withdrew his permission for me to record our interview as soon as we finished talking, which of course I did, and then he did not respond to my follow-up interview requests. I have a greater understanding of this reluctance after following this cohort over the course of layoffs. To protect their anonymity, I changed the names and modified identifying details for all of the respondents.

The interviews for this study were conducted by phone. Although face-to-face interviews are preferable to phone interviews, they were impractical in this case since the sample is widely dispersed throughout the United States and abroad.[3] By the time of the phone interviews, respondents already had filled in three surveys, so I knew a great deal of background information about each respondent, including their demographic characteristics, salary, and performance ratings. In the interviews, I asked them to reflect and elaborate on their reasons for entering the industry, their work experiences, and their career aspirations. Questions covered the following topics: choice of college degree; job search; past and present job responsibilities; experience with corporate retention programs; relationships with coworkers and supervisors; and future plans. My interview schedule included follow-up questions related to each respondent's specific answers to the survey. For instance: "In your survey, you indicated that there is a ____ percent chance that you will be at your current employer in the next ten years. Do you recall why you wrote that?" Interviewees were also asked to clarify comments they wrote in response to the survey's open-ended questions. For example, one engineer wrote on the time 1 survey that he was "dissatisfied with current position," while at time 3 he wrote "very positive view of company." In the in-depth interview, I asked him to speculate why his view had changed.

I followed up with this original group two more times, in 2015 and 2016. Of the 44 people in my original sample, 23 stayed with me until the end. Among the geoscientists, 15 of the original 22 stayed on. My sample does not include any men who were dismissed during the downturn. This could have been the result of chance, or it could reflect a gendered reluctance to discuss hardships at work. It is possible that people dropped out of the study because they did not want to talk to me about the circumstances of their departures. I have missing information on four of the cases I followed (see table A1). I know from their LinkedIn profiles that two geoscientists (Brian and Sylvia) moved to other oil companies, but I do not know the circumstances of their departures from GOG because they refused my repeated requests for follow-up interviews.

Table A1 shows the total number of surveys and interviews conducted for this study. I conducted 43 interviews at time 3, 32 at time 4, and 23 at time 5, for a total of 98 interviews (see table A1). Interviews lasted between thirty minutes and an hour. I transcribed all of the interviews myself. Transcribing is tedious work, but it enables me to grasp and to process what respondents are saying and what they are

Total Interview Sample
Surveys and Interviews Completed

Gender and Occupation	Surveys Only		Surveys and In-Depth Interviews			GOG Employment Status*
	T1	T2	T3	T4	T5	
Women Engineers	x	x	x			Leaver
	x	x	x			Leaver
	x	x	x	x		Stayer
	x	x	x	x	x	Stayer
	x	x	x			no info
Women Geoscientists	x	x	x	x	x	Leaver
	x	x	x		x	Leaver
	x	x	x	x	x	Leaver
	x	x	x	x	x	Leaver
	x	x	x		x	Leaver
	x	x	x	x	x	Leaver
	x	x	x	x		Leaver
	x	x	x	x		Leaver
	x	x	x	x		Leaver
	x	x	x	x	x	Leaver
	x	x	x	x	x	Leaver
	x	x	x	x	x	Stayer
	x	x	x	x	x	Stayer
	x	x	x	x	x	Stayer
Men Engineers	x	x	x	x	x	Leaver
	x	x	x	x	x	Leaver
		x	x	x		Leaver
	x	x	x			Leaver
	x	x	x	x		Leaver
	x	x	x	x		Leaver
	x		survey only	x		Leaver
	x	x	x	x	x	Stayer
	x	x	x	x	x	Stayer
	x	x	x	x	x	Stayer

TABLE AI

(continued)

Gender and Occupation	Surveys Only		Surveys and In-Depth Interviews			GOG Employment Status*
	T1	T2	T3	T4	T5	
	x	x	x	x	x	Stayer
	x	x	x	x		Stayer
	x	x	x	x	x	Stayer
	x	x	x	x	x	Stayer
	x	x	x			*no info*
	x	x	x			*no info*
	x	x				*no info*
Men Geoscientists	x	x	x			Leaver
	x	x	x	x	x	Leaver
	x	x	x			Leaver
	x	x	x	x	x	Leaver
	x	x	x	x	x	Stayer
	x	x	x	x	x	Stayer
	x	x	x	x		Stayer
	x	x	x	x	x	Stayer

* At the end of the study period

not saying. The pauses, the gaps, the feelings expressed—all of these make indelible impressions over the many hours spent transcribing.

Although I interviewed 44 scientists and engineers over the years of this study, this book focuses on my subsample of 22 geoscientists (see table A2). There are many reasons for this decision. Industry engineers and scientists have distinct backgrounds and career trajectories. Engineers enter the industry with a bachelor's degree, while geoscientists have a graduate degree. Unlike geoscience education, college training prepares engineers for working in industry, including for management

Geoscientist Interview Sample

Gender	Pseudonym	GOG Employment Status*	Family Status*		Current Position
			Married	*Kids*	
Women	Alice	Leaver	yes	yes	unemployed
	Barb	Leaver	yes	yes	vet school
	Claire	Leaver	yes	yes	teacher
	Elena	Leaver	yes	yes	new co.
	Eve	Stayer	yes	no	GOG
	Gloria	Leaver	no	no	new co.
	Irene	Leaver	yes	no	new co.
	Karen	Leaver	yes	no	consultant
	Kimberly	Leaver	no	no	new co.
	Leticia	Leaver	no	no	new co.
	Lynne	Stayer	yes	no	GOG
	Mary	Leaver	yes	yes	PhD program
	Patricia	Stayer	yes	no	GOG
	Sylvia	Leaver	yes	yes	new co.
Men	Ben	Stayer	yes	yes	GOG
	Brian	Leaver	no	no	new co.
	Colt	Leaver	yes	no	new co.
	Jordan	Stayer	yes	yes	GOG
	Martin	Stayer	yes	yes	GOG
	Sam	Stayer	yes	yes	GOG
	Steven	Leaver	no	no	new co.
	William	Leaver	yes	no	new co.

* Employment and family status at the end of the study period

positions. (Most oil and gas companies are headed by engineers or people with finance backgrounds.) Engineers are closer to the drill bit than geoscientists are, often living and working on site (instead of in downtown offices); many also work on rotation schedules (e.g., three weeks on/three weeks off). Engineers have many more employment options than geoscientists have, including in several industries unrelated to oil and gas (for example, one came to the industry after working for an alcohol distillery). Regarding gender, women make up a smaller percentage of engineers compared to geoscientists. At GOG, only five women engineers volunteered to participate in the interview portion of the study, and all but one eventually dropped out. I'm not sure why that happened, but their lack of participation prevented me from exploring company dynamics through their experiences.

WHY CONDUCT IN-DEPTH INTERVIEWS?

Sociologists who use in-depth interviews believe that people's thoughts and feelings matter in our sociological explanations. This is tied to basic assumptions about the social world. Instead of seeing people as vessels of culture or social structure, we see them as active agents involved in interpreting, navigating, and strategizing in their social settings. The goal of the in-depth interview is to understand why people do what they do from their points of view. The only way we can get at this information is by talking to them.

Veteran in-depth interviewer Allison Pugh (2013) acknowledges that this method gets a bad rap from some cultural sociologists who consider interviews to be merely ex-post facto rationalizations or justifications for behavior. Detractors charge that since respondents will always attempt to portray themselves in an honorable light, they inevitably produce slanted accounts of their social lives that have dubious truth-value.

Pugh agrees that this "honorable display" is part of what one learns in an in-depth interview. But this is not the whole story. Successful interviewers probe beneath these surface accounts. She describes three additional levels of understanding that can emerge from in-depth interviews: the schematic (referring to the frameworks or general worldviews our respondents embrace), the visceral (moral sentiments that provoke physical reactions, including disgust, longing, and despair), and the metafeelings (their assessments of how they should have reacted in the social situations they describe). At each of these levels, respondents can express ambivalence—a universal human experience that some other methods cannot even access.

My epistemological approach is best described as "socioanalysis," a term that Pierre Bourdieu uses to describe the methodology of sociology. In Bourdieu's formulation, socioanalysis is akin to psychoanalysis. Like a psychoanalyst, a sociologist attempts to uncover the hidden structures of a society embedded in individuals' taken-for-granted assumptions about social life. Describing this approach, Loïc Wacquant explains:

> "Actors" are always structurally constituted, each habitus being an individuated permutation of social structures internalized in the form of cognitive and evaluative categories. (1990:685)

Bourdieu-ian sociology uncovers the socialized interests that are internalized in each and every one of us. This is precisely the value of in-depth interviewing, in my view. By talking to people, a sociologist can identify the cognitive and evaluative categories they use, draw connections between those categories and the reproduction of power and inequality, and pave the way for a new orthodoxy that is less oppressively grounded in capitalism, patriarchy, and racism.

In my case, I conducted interviews with scientists and engineers at GOG to identify what this corporation was or was not doing to promote diversity and the retention of women scientists. Thus, I was interviewing individuals about their career decisions, but my focus was on the organization. By delving into the viewpoints and experiences of scientists at GOG, my goal was to learn how this particular company structures opportunities for scientists with the goal of explaining its continued white male domination. My respondents could not directly answer this question for me, but they could offer personal accounts about their career trajectories, and from these accounts I was able to make inferences about the structure of the organization.

My approach also draws from methodological principles set forth by C. Wright Mills. In his classic formulation of the "sociological imagination," Mills (1959) writes of how personal struggles reflect the workings of larger social forces that may be invisible to those on the ground, but that shape the conditions of their existence. The sociologist's role is to investigate, document, and critique these forces.

Dorothy Smith (1979) takes this mantle further, arguing that a "sociology for women" should investigate not only the problems faced by individuals in their everyday lives, but the categories of analysis used by sociologists to understand their reality. Both can reflect the process of domination. Sociology and other kinds of expert knowledge are riddled with sexism, she says. To fight against these oppressive frameworks, sociologists should start their studies looking at the world from the vantage point of individual women, investigating how women understand their lives and how they make choices within the contexts of their everyday lives. Thus she urges gender scholars to ask, how do women solve the practical problems they face? Starting with this question will make intelligible what might seem like irrational choices from elite men's vantage points. For example, why do women leave their lucrative careers in science? All sorts of myths obscure sociological understanding of this problematic, including the notion that women's choices are "for the family," a sexist canard that sociologist Sarah Damaske (2011) masterfully debunked in her study of that name. To make sense of women's decisions requires understanding how women see their own lives and their choices, and not assuming that women are essentially different from and inferior to men. In-depth interviews are exceptionally well suited to uncover these logics and "problematics" (using Dorothy Smith's language) that are confronted by individuals in their daily lives. Interviews give respondents the opportunity to reflect on their lives and make sense of their choices.

In addition to conducting and analyzing these interviews, I attended numerous talks, seminars, and conventions focused on diversity and inclusion in the oil industry. I was an invited speaker at several of these, including at the annual meetings of the AAPG and the American Geophysical Union (AGU), where I became an unpaid consultant on their sexual harassment policy. My affiliation with the University of Texas at

Austin gave me access to several campus events organized through the Kay Bailey Hutchison Center for Energy, Law, and Business. I talked to countless individuals at these venues who contributed to my understanding of the oil and gas industry. My participation in these events enhanced my analysis of how the industry is organized and how it is positioned in the larger field of energy production.

ANALYSIS

This book mostly focuses on the career narratives of individual scientists, which is not the typical way that sociologists analyze interview data. More commonly, sociologists identify themes that emerge from rereading the transcripts and then analyze those themes as discursive constructions of hegemonic ideas and practices. That is how I wrote chapter 2, on the motivations of my respondents to study geology and to work for the oil and gas industry. But the technique seemed ill suited for the subsequent chapters. Chopping up the scientists' narratives into themes seemed to diminish what made these accounts special—individuals' efforts over time to make sense of and respond to their changing work conditions. That is why I switched the focus to individual career narratives.

This analytic approach is akin to the "individuology" advocated by sociologist and psychoanalyst Nancy Chodorow (2020). With her unique background straddling these two disciplines, Chodorow understands the value of both looking for patterns and respecting individual difference. She writes,

> You can know a lot about someone when you know their race, family, ethnicity, gender, and social and historical location....Yet however fine-tuned this sociocultural and intersectional analysis, you can never predict individual subjectivity. People bring emotion, temperament, fantasy, and wish ... to whatever comes from without. (260)

Chodorow reminds sociologists that not everyone responds similarly to the same environment or even to the same traumatic experience. The same is true for women scientists employed at GOG. I hope that by recounting their personal narratives about their careers, I honor their agency and creativity in responding to the structural constraints they experienced in their organizational environment.

Very little is written about analyzing longitudinal interview data. One exception is a helpful article by sociologist Joseph Hermanowicz (2013) in which he lays out the advantages of conducting longitudinal qualitative interviews. Hermanowicz studied academic scientists at a ten-year interval with the goal of uncovering how people experience, interpret, and respond to changes over the course of their career. He wanted to document "academics' shifting perceptions of their jobs to uncover the meanings they invest in their work, when and where they find satisfaction, how they succeed and fail, and how the rhythms of work change as they age." Doing so allowed Hermanowicz to thematize the preoccupations of scientists at different stages in their academic careers.

In my study, I am less interested in understanding how people change, and more interested in their changing circumstances. My respondents are acute observers of a world that is otherwise inaccessible to me. Their impressions at different points in time help me to understand how their oil company responded to the downturn in prices, and the toll that the company's downsizing strategy exacted on their geotechnical workforce.

I took inspiration for my approach from a number of books, including Arlie Hochschild's *Second Shift* (1989), Judith Stacey's *Brave New Families* (1998), and Aliya Rao's *Crunch Time* (2020). These studies reveal the workings of a patriarchal society through the personal narratives of individuals coping with daily challenges in their families. I hope that my decision to focus on the careers of individual geoscientists captures these dynamics in the oil and gas industry.

NOTES

CHAPTER I. GENDER, GEOLOGY, AND THE OIL AND GAS INDUSTRY

1. www.aapg.org/career/aapg-net/about/articleid/38966/groupid/949 /aff/330, Sept. 26, 2018. PROWESS has since changed its name to the AAPG Women's Network and altered its mission statement.

2. I do not mean to imply that older forms of work organization were better at achieving diversity. During the heyday of Fordism, when stable employment was more of the norm than it is today, good jobs typically excluded women and minority men. During economic downturns, members of these groups who managed to gain access to these good jobs were among the "first fired" due to seniority rules (Lichtenstein 2002).

3. www.ran.org/the-understory/don_t_take_our_word_for_it _chevron_s_we_agree_campaign_one_of_2010_s_worst/.

4. Women are also missing from Yergin's follow-up volume, *The Quest* (2011). Lest this be considered a failure only of male authors, recent books by Naomi Klein (2014) and Rachel Maddow (2019) also feature only male actors in their analyses of the oil industry. For an alternative feminist perspective, see Bell, Daggett, and Labuski 2020.

5. "Land man" is a job relating to researching property titles and leasing oil rights. "Company man" refers to the onsite corporate

representative who manages a well site that is drilled by a subcontractor or corporate partner.

6. Very little evidence supports this assumption. See Kalev, Kelly, and Dobbin 2006; Kaiser et al. 2013; Castilla and Benard 2010.

CHAPTER 2. THE OIL AND GAS PIPELINE

1. One study of doctoral recipients in the 1990s found high rates of unemployment, with geologists and environmental scientists among the most likely to be unemployed (Shettle 1997).

2. Goldie Blox: www.goldieblox.com/. Legos produced a "science lab" for girls designed by a woman geoscientist: www.nytimes.com /2014/08/22/business/short-lived-science-line-from-lego-for-girls.html. See also Thébaud and Charles 2018.

3. The metaphor of a "pipeline" is rejected by some feminists because they think it portrays women as passive victims of an inflexible structure, and because it ignores the ways that the obstacles to and opportunities for success may vary for different groups of women and minority men. See Branch 2016.

4. She Can Stem website: https://shecanstem.com/. The Society for Women Engineers: https://swe.org/wp-content/uploads/2018/12 /Invent-it-Build-it-WE18-Press-Release-FINAL.pdf. Google's program: https://www.madewithcode.com/about/.

5. https://thebeardedladyproject.com/portraits/.

6. This is according to the website hbcu-colleges.com based on information from 2018.

7. Website of the American Association of Blacks in Energy: www .aabe.org/. For information about the NAACP environmental policy: https://naacp.org/environmental-climate-justice-about/.

8. Ash's interview is available here: https://therealnews.com/stories /mash0902peri.

CHAPTER 3. THE STAYERS

1. According to sociologist Jennifer Glass and her colleagues (Glass et al. 2013), 31.5 percent of STEM-educated women leave (compared to

6 percent of women in other professions who leave); those with more advanced STEM degrees leave more frequently than do those with fewer degrees. Marriage also makes a difference: Compared to other women professionals, STEM women are 84 percent more likely to leave when they get married, unless they have a spouse who is also in STEM, in which case they are less likely to leave. Importantly, those who leave do not drop out of the labor force, but rather move into non-STEM occupations.

2. Two of my favorite autobiographies of successful women scientists are botanist Hope Jarhren's *Lab Girl* (2016) and physicist Eileen Pollack's *The Only Woman in the Room* (2015).

3. My paranoia about this possible outcome was stoked at the 2020 Women's Global Leadership Conference, where a Schlumberger official touted the benefits of using "data analytics" tools for hiring, based on "a broad range of characteristics on candidate profiles, including demographics, academic and behavioral attributes." The goal is to "provide information back to hiring managers." Although these tools were pitched as a means to address unconscious bias, the spokesperson assured the audience that "the tactic does not eliminate candidates we would have typically selected."

4. "Geo" is a common shorthand expression for "geoscientist," a term that encompasses geologists, geophysicists, and petrophysicists.

5. Women's Global Leadership Conference in Energy, Houston, Texas, October 28–29, 2019.

6. As Allison Pugh (2013:10) observes, in-depth interviews enable respondents to express emotional ambivalence through gestures like laughing, sighing, and stuttering. In effect, respondents "'tell' us what kind of things are uncomfortable, horrifying, emboldening, joyful, and they do so through non-verbal means that communicate their emotional frameworks." In contrast, survey research excludes this layer of meaning from analysis.

7. One told me that he did not put his PhD credential on his business card, and he suspected that many of his coworkers were not aware that he had a doctoral degree.

8. Studies that rely on volunteer respondents can unintentionally omit entire groups who opt not to participate, potentially skewing the results. See the methodological appendix.

9. Sam's selection for a job transfer may be another hidden advantage for white workers, a suggestion made by Adia Harvey Wingfield (private communication). In general, Black and Latinx people—who are virtually absent from my sample—move more often than white people do, but they may be perceived by managers as less able or amenable to accepting a transfer to another state. She suggests that this is another way that the ideal worker norm can favor white men.

CHAPTER 4. VOLUNTARY SEPARATIONS

1. Looze (2017) contends that white mothers are especially inclined to remain with their employer for this reason.

2. Steven had a lot of critical things to say about GOG once the layoffs began, which I discuss in the conclusion to this chapter.

3. I lack information about two of those who transferred after the crash (one man and one woman), who declined multiple requests for a follow-up interview. I do not know if their reasons for leaving were voluntary or involuntary, but according to their LinkedIn profiles, both have jobs in other oil companies.

CHAPTER 5. CORPORATE DOWNSIZING

1. 350,000 jobs lost (https://oilprice.com/Energy/Energy-General /After-350000-Layoffs-Oil-Companies-Now-Face-Worker-Shortages .html), June 8, 2016. 258,000 jobs lost, end of 2015, concentrated in E and P (upstream), www.usatoday.com/story/money/2015/12/31/energy -oil-gas-us-jobs/78066610/.

2. I am not sure why my sample does not include any men who left during this period. See methodological appendix.

3. I do not discuss the case of the forced retirement out of concerns regarding deductive disclosure.

4. "FMLA" refers to the US Family and Medical Leave Act, which guarantees twelve weeks of unpaid leave with job protection to most workers at large establishments.

CHAPTER 6. ORGANIZATIONAL GASLIGHTING

1. www.nytimes.com/2020/04/21/business/energy-environment/coronavirus-oil-prices-collapse.html; www.statesman.com/business/20200505/texas-oil-and-gas-regulators-reject-production-cuts.

2. Movie examples include *Temple Grandin* (2010), about a woman scientist's efforts to bring humane treatment to the livestock industry, and *Gorillas in the Mist* (1988), about Dian Fossey's efforts to protect the endangered African mountain gorilla. The movie *Hidden Figures* (2016), about NASA scientists Katherine Johnson, Dorothy Vaughan, and Mary Jackson, who worked for the astronaut mission, is a rare example that celebrates the contributions of Black women scientists.

3. Because of the male domination of the oil and gas industry, some are optimistic that women are poised to make inroads into the burgeoning renewable energy industry. At the Women's Global Leadership Conference 2020, a spokesman from Schlumberger maintained that since women lack a "legacy" in oil and gas, they are less likely than men are to resist the energy transition, and are ideally situated to take up leadership roles in new wind and solar energy companies. See also Allison, McCrory, and Oxnevad 2019.

METHODOLOGICAL APPENDIX

1. See American Association of Petroleum Geologists 2009; https://explorer.aapg.org/story/articleid/1819/itaposs-a-tough-place-for-a-woman.

2. We received helpful feedback on our proposal from the NSF reviewers, but ultimately we sensed an unwillingness on the part of NSF to fund a study of the oil industry, which has ample resources to conduct their own research. Chandra and I decided early on not to accept research funding from the oil industry.

3. I prefer phone calls to video calls. At the time of these interviews, the Zoom call was not yet part of the common lexicon. The pandemic has made me more adept at video calls, but I still prefer phone calls because I find them less distracting and performative, but this view is not shared by other qualitative researchers. See "Qualitative and Quarantined: Techniques and Ethics of Online Interview Research," recorded webinar hosted by the American Sociological Association on September 22, 2020.

REFERENCES

Acker, Joan. 1990. Hierarchies, Jobs, Bodies: A Theory of Gendered Organizations. *Gender and Society* 4:139–58.

ADVANCEGeo. 2019. "Empowering Geoscientists to Transform Workplace Culture." July 24, 2020. https://serc.carleton.edu /advancegeo/index.html.

Alfrey, Lauren, and France Winddance Twine. 2017. "Gender Fluid Geek Girls: Negotiating Inequality Regimes in the Technology Industry." *Gender and Society* 31(1):28–50.

Allison, Juliann Emmons, Kirin McCrory, and Ian Oxnevad. 2019. "Closing the Renewable Energy Gender Gap in the United States and Canada: The Role of Women's Professional Networking." *Energy Research and Social Science* 55:35–45.

American Association of Blacks in Energy. 1998. "The History of the American Association of Blacks in Energy." Original Written by Rufus W. McKinney—May 1994, Bethesda, MD. Edited by Robert L. Hill—August 1998, Washington, DC. Downloaded June 10, 2019. www.aabe.org/index.php?component=pages&id=14.

American Association of Petroleum Geologists (AAPG). 2009. "Results from the American Association of Petroleum Geologists (AAPG) Professional Women in the Earth Sciences (PROWESS)

Survey." June 7, 2011. www.aapg.org/committees/prowess/AAPG
_Jun3.final.pdf.

Ash, Michael, and James Boyce. 2018. "Racial Disparities in Pollution Exposure and Employment at US Industrial Facilities." *PNAS* (Proceedings of the National Academy of Sciences) 115(42). www .pnas.org/cgi/doi/10.1073/pnas.1721640115.

Ashcraft, Karen Lee. 2013. "The Glass Slipper: 'Incorporating' Occupational Identity in Management Studies." *Academy Of Management Review* 38(1):6–31.

Babcock, Linda, and Sara Laschever. 2003. *Women Don't Ask: Negotiation and the Gender Divide*. Princeton, NJ: Princeton University Press.

Bansak, Cynthia, and Steven Raphael. 2008. "The State Children's Health Insurance Program and Job Mobility: Identifying Job Lock among Working Parents in Near-Poor Households." *ILR Review* 61(4). https://digitalcommons.ilr.cornell.edu/ilrreview/vol61/iss4/7/.

Beamish, Thomas. 2002. *Silent Spill: The Organization of an Industrial Crisis*. Cambridge, MA: MIT Press.

Bell, E.E., and S. Nkomo. 2001. *Our Separate Ways: Black and White Women and the Struggle for Professional Identity*. Boston: Harvard Business School Press.

Bell, Shannon Elizabeth, Cara Daggett, and Christine Labuski. 2020. "Toward Feminist Energy Systems: Why Adding Women and Solar Panels Is Not Enough." *Energy Research and Social Science* 68:1–13.

Bell, Shannon Elizabeth, Jenrose Fitzgerald, and Richard York. 2019. "Protecting the Power to Pollute: Identity Co-Optation, Gender, and the Public Relations Strategies of Fossil Fuel Industries in the United States." *Environmental Sociology* 5:1–16.

Benard, Stephen, and Shelley Correll. 2010. "Normative Discrimination and the Motherhood Penalty." *Gender and Society* 24:616–46.

Bernard, Rachel, and Emily Cooperdock. 2018. "No Progress on Diversity in 40 Years." *Nature Geoscience* 11:292–95.

Berrey, Ellen. 2015. *The Enigma of Diversity*. Chicago: University of Chicago Press.

Bosky, Amanda, Chandra Muller, and Christine Williams. 2017. "The Price of Crude and the Deserving Professional: Gender Inequality in the Oil Industry." Paper presented at the annual meetings of the American Sociological Association, Montreal, August.

Branch, Enobong Hannah. 2016. *Pathways, Potholes, and the Persistence of Women in Science: Reconsidering the Pipeline.* Lanham, MD: Lexington.

Brand, Jennie F. 2015. "The Far-Reaching Impact of Job Loss and Unemployment." *Annual Review of Sociology* 41:359–75.

Britton, Dana. 2017. "Beyond the Chilly Climate: The Salience of Gender in Women's Academic Careers." *Gender and Society* 31(1):5–27. doi: 10.1177/0891243216681494.

Britton, Dana, and Laura Logan. 2008. Gendered Organizations: Progress and Prospects. *Sociology Compass* 2:107–21.

Brockner, Joel. 1992. "Managing the Effects of Layoffs on Survivors." *California Management Review* 34(2):9–28.

Bullard, Robert D. 2019. "African Americans on the Frontline Fighting for Environmental Justice." https://drrobertbullard.com/african-americans-on-the-frontline-fighting-for-environmental-justice/.

Burek, C.V., and B. Higgs, eds. 2007. *The Role of Women in the History of Geology.* Special Publications 281. London: Geological Society.

Byrne, Jason, and Jennifer Wolch. 2009. "Nature, Race, and Parks: Past Research and Future Directions for Geographic Research." *Progress in Human Geography* 33(6):743–65.

Byron, Reginald. 2010. "Discrimination, Complexity, and the Public /Private Sector Question." *Work and Occupations* 37(4):435–75.

Cai, Zhengyu, Karen Maguire, and John Winters. 2019. "Who Benefits from Local Oil and Gas Employment? Labor Market Composition in the Oil and Gas Industry in Texas." IZA-Institute of Labor Economics, Discussion Paper series #12349. http://ftp.iza.org/dp12349.pdf.

Castilla, Emilio, and Stephen Benard. 2010. "The Paradox of Meritocracy in Organizations." *Administrative Science Quarterly* 55:543–76.

Catalyst. 2019. "Women in Energy—Gas, Mining, and Oil: Quick Take." June 22, 2020. www.catalyst.org/research/women-in-energy-gas-mining-oil/.

Cha, Youngjoo. 2014. "Job Mobility and the Great Recession: Wage Consequences by Gender and Parenthood." *Sociological Science* 1:159–77.

Charrad, Mounira. 2009. "Kinship, Islam, or Oil: Culprits of Gender Inequality?" *Politics and Gender* 5(4):546–53.

Chodorow, Nancy. 2020. *The Psychoanalytic Ear and the Sociological Eye.* New York: Routledge.

Clancy, Katheryn, Robin G. Nelson, Julienne N. Rutherford, and Katie Hinde. 2014. "Survey of Academic Field Experiences (SAFE): Trainees Report Harassment and Assault." *PLOS-One.* https://doi.org/10.1371/journal.pone.0102172.

Collins, Caitlyn. 2019. *Making Motherhood Work.* Princeton, NJ: Princeton University Press.

Collins, Patricia Hill. 1999. "Moving beyond Gender: Intersectionality and Scientific Knowledge." Pp. 261–84 in *Revisioning Gender,* edited by Myra Marx Ferree, Judith Lorber, and Beth Hess. Thousand Oaks, CA: Sage.

Collins, Sharon. 2011. "From Affirmative Action to Diversity: Erasing Inequality from Organizational Responsibility." *Critical Sociology* 37:517–20.

Connell, R. W. 1998. "Masculinities and Globalization." *Men and masculinities* 1(1):3–23.

Correll, Shelley. 2001. "Gender and the Career Choice Process: The Role of Biased Self-Assessments." *American Journal of Sociology* 106(6):1691–730.

Correll, Shelley, Stephen Benard, and In Paik. 2007. "Getting a Job: Is There a Motherhood Penalty?" *American Journal of Sociology* 112(5): 1297–338.

Correll, Shelley, Erin Kelly, Lindsey Trimble O'Connor, and Joan Williams. 2014. "Redesigning, Redefining Work." *Work and Occupations* 41(1):3–17.

Daggett, Cara. 2018. "Petro-Masculinity: Fossil Fuels and Authoritarian Desire." *Millennium: Journal of International Studies* 47(1):25–44.

Damaske, Sarah. 2011. *For the Family? How Class and Gender Shape Women's Work.* New York: Oxford University Press.

DiMaggio, Paul. 2001. *The Twenty-First Century Firm.* Princeton, NJ: Princeton University Press.

Dobbin, Frank. 2009. *Inventing Equal Opportunity.* Princeton, NJ: Princeton University Press.

Dobbin, Frank, Alexandra Kalev, and Erin Kelly. 2007. "Diversity Management in Corporate America." *Contexts* 6(4):21–27.

Dobbin, Frank, S. Kim, and Alexandra Kalev. 2011. "You Can't Always Get What You Need: Organizational Determinants of Diversity Programs." *American Sociological Review* 76:386–411.

England, Paula. 2010. "The Gender Revolution: Uneven and Stalled." *Gender and Society* 24:149–66.

Enloe, Cynthia. 2014. *Seriously! Investigating Crashes and Crises as if Women Mattered.* Berkeley: University of California Press.

Ervin, Brian. 2016. "Salary Survey Points to Experience Gap." *AAPG Explorer,* June, p. 6.

Extractive Industries for Development Series #28. http://documents .worldbank.org/curated/en/266311468161347063/Extracting-lessons-on -gender-in-the-oil-and-gas-sector.

Finney, Carolyn. 2014. *Black Faces, White Spaces: Reimagining the Relationship of African Americans to the Great Outdoors.* Chapel Hill: University of North Carolina Press.

Fouad, Nadya A., Romila Singh, Kevin Cappaert, Wen-hsin Chang, and Min Wan. 2016. "Comparison of Women Engineers Who Persist in or Depart from Engineering." *Journal of Vocational Behavior* 92:79–93.

Friedman, Barry. 2016. "Don't Call Her a 'Woman Geoscientist.'" *AAPG Explorer.* https://explorer.aapg.org/story/articleid/28934/dont -call-her-a-woman-geoscientist.

Gabriel, Yiannis, David E. Gray, and Harshita Goregaokar. 2010. "Temporary Derailment or the End of the Line? Managers Coping with Unemployment at 50." *Organization Studies* 31(12): 1687–712.

Gabriel, Yiannis, David E. Gray and Harshita Goregaokar. 2013. "Job Loss and Its Aftermath among Managers and Professionals: Wounded, Fragmented and Flexible." *Work, Employment and Society* 27(1):56–72.

Giordano, Sara. 2017. "Those Who Can't, Teach: Critical Science Literacy as a Queer Science of Failure." *Catalyst: Feminism, Theory, Technoscience* 3(1):1–21.

Glass, Jennifer. 2009. "Work-Life Policies: Future Directions for Research." Pp. 231–50 in *Work Life Policies That Make a Difference,* edited by Alan Booth and Nan Crouter. New York: Russell Sage.

Glass, Jennifer, and Mary Noonan. 2016. "Telecommuting and Earnings Trajectories among American Women and Men, 1989–2008." *Social Forces* 95(1):217–50.

Glass, Jennifer, Sharon Sassler, Yael Levitte, and Kathering M. Michelmore. 2013. "What's So Special about STEM? A Comparison of Women's Retention in STEM and Professional Occupations." *Social Forces* 92(2):723–56.

Gorman, Elizabeth, and Sarah Mosseri. 2019. "How Organizational Characteristics Shape Gender Difference and Inequality at Work." *Sociology Compass* 13:1–18.

Gornick, Janet, and Marcia Meyers. 2003. *Families That Work: Policies for Reconciling Parenthood and Employment*. New York: Russell Sage Foundation.

Gries, Robbie Rice. 2017. *Anomalies: Pioneering Women in Petroleum Geology: 1917–2017*. Denver: JeWel.

Hanson, Rebecca, and Patricia Richards. 2019. *Harassed: Gender, Bodies, and Ethnographic Research*. Oakland: University of California Press.

Harris, Cheryl. 2019. "Quitting Science: Factors That Influence Exit from the STEM Workforce." *Journal of Women and Minorities in Science and Engineering* 25(2):93–118.

Heilman, Madeline E., and Michelle C. Haynes. 2005. "No Credit Where Credit Is Due: Attributional Rationalization of Women's Success in Male–Female Teams." *Journal of Applied Psychology* 90(5):905–16.

Herman, Clem, Suzan Lewis, and Anne Laure Humbert. 2013. "Women Scientists and Engineers in European Companies: Putting Motherhood under the Microscope." *Gender, Work, and Organization* 20:467–78.

Hermanowicz, Joseph. 2013. "The Longitudinal Qualitative Interview." *Qualitative Sociology* 36:189–208.

Hewlett, Sylvia A. 2007. *Off-ramps and On-ramps: Keeping Talented Women on the Road to Success*. Boston: Harvard Business School Publishing.

Hill, Patricia Wonch, Julia McQuillan, Eli Talbert, Amy Spiegel, G. Robin Gauthier, and Judy Diamond. 2017. "Science Possible Selves and the Desire to Be a Scientist: Mindsets, Gender Bias, and Confidence during Early Adolescence." *Social Sciences* 6(55). doi:10.3390/socsci6020055.

Hiltzik, Michael. 2015. "Tech Industry's Persistent Claim of Worker Shortage May Be Phony." *Los Angeles Times*, August 1. www.latimes.com/business/hiltzik/la-fi-hiltzik-20150802-column.html.

Hira, R. 2010. "U.S. Policy and the STEM Workforce System." *American Behavioral Scientist* 53:949–61.

Hirschman, Albert. 1970. *Exit, Voice, and Loyalty: Responses to Decline in Firms, Organizations, and States.* Cambridge, MA: Harvard University Press.

Hirsh, Elizabeth, and Donald Tomaskovic-Devey. 2020. "Metrics, Accountability, and Transparency: A Simple Recipe to Increase Diversity and Reduce Bias." Pp. 16–23 in *What Works? Evidence-based Ideas to Increase Diversity, Equity, and Inclusion in the Workplace.* University of Massachusetts, Center for Employment Equity. www.umass.edu/employmentequity/what-works.

Hochschild, Arlie, with Anne Machung. 1989. *The Second Shift.* New York: Viking.

Hollister, Matissa, and Kristin Smith. 2014. "Unmasking the Conflicting Trends in Job Tenure by Gender in the United States, 1983–2008." *American Sociological Review* 79(1):159–81.

Holmes, Mary Anne, Suzanne OConnell, and Kuheli Dutt. 2015. *Women in the Geosciences: Practical, Positive Practices toward Parity.* Washington, DC: American Geophysical Union (in conjunction with John Wiley and Sons).

Hunt, Jennifer. 2016. "Why Do Women Leave Science and Engineering?" *ILR Review* 69(1):199–226.

Huntoon, Jacqueline E., and Melissa J. Lane. 2007. "Diversity in the Geosciences and Successful Strategies for Increasing Diversity." *Journal of Geoscience Education* 55(6):447–57. doi: 10.5408/1089-9995-55.6.447.

IHS. 2014. *Minority and Female Employment in the Oil and Gas and Petrochemical Industries.* Report prepared for the American Petroleum Institute. Washington, DC: IHS Global.

Jacobs, Jerry A., Seher Ahmad, and Linda J. Sax. 2017. "Planning a Career in Engineering: Parental Effects on Sons and Daughters." *Social Sciences* 6(2). doi:10.3390/socsci6010002.

Jahren, Hope. 2016. *Lab Girl.* New York: Random House.

Johnson, Angela C. 2007. "Unintended Consequences: How Science Professors Discourage Women of Color." *Science Education* 91(5):805–21.

Jung, J. 2017. "A Struggle on Two Fronts: Labor Resistance to Changing Layoff Policies at Large U.S. Companies." *Socio-Economic Review* 15:213–39.

Kaiser, Cheryl R., Brenda Major, Ines Jurcevic, Tessa Dover, Laura Brady, and Janessa Shapiro. 2013. "Presumed Fair: Ironic Effects of Organizational Diversity Structures." *Journal of Personality and Social Psychology* 104(3):504–19.

Kalev, Alexandra. 2014. "Who You Downsize Is How You Downsize: Biased Formalization, Accountability and Managerial Diversity." *American Sociological Review* 79(1):109–35.

Kalev, Alexandra. 2020. "Research: U.S. Unemployment Rising Faster for Women and People of Color." *Harvard Business Review*, April 20. https://hbr.org/2020/04/research-u-s-unemployment-rising-faster-for-women-and-people-of-color.

Kalev, Alexandra, Erin Kelly, and Frank Dobbin. 2006. "Best Practices or Best Guesses? Assessing the Efficacy of Corporate Affirmative Action and Diversity Policies." *American Sociological Review* 71(4):589–617.

Kalleberg, Arne. 2011. *Good Jobs, Bad Jobs: The Rise of Polarized and Precarious Employment Systems in the United States, 1970s-2000s.* New York: Russell Sage Foundation.

Kalleberg, Arne. 2018. *Precarious Lives: Job Insecurity and Well-Being in Rich Democracies.* Cambridge: Polity.

Kanter, Rosabeth Moss. 1977. *Men and Women of the Corporation.* New York: Basic.

Karamessini, Maria, and Jill Rubery. 2014. *Women and Austerity.* London and New York: Routledge.

Kelly, Erin, and Phyllis Moen. 2020. *Overload: How Good Jobs Went Bad and What We Can Do about It.* Princeton: Princeton University Press.

Kilanski, Kristine. 2011. "Gender, Graduate School, and the Geosciences." MA thesis, University of Texas at Austin.

Kirsch, Stuart. 2010. "Sustainability and the BP Oil Spill." *Dialectical Anthropology* 34(3):295–300.

Kirsch, Stuart. 2014. *Mining Capitalism: The Relationship between Corporations and Their Critics.* Oakland: University of California Press.

Klein, Naomi. 2014. *This Changes Everything: Capitalism vs. the Climate.* New York: Simon and Schuster.

Koeber, Charles. 2002. "Corporate Restructuring, Downsizing, and the Middle Class: The Process and Meaning of Worker Displacement in the 'New' Economy." *Qualitative Sociology* 25(2):217–46.

Lane, Carrie M. 2011. *A Company of One: Insecurity, Independence and the New World of White-Collar Employment.* Ithaca, NY: Cornell University Press.

Lee, Sophia, and Asima Ahmad. 2019. "Understanding Unconscious Bias Is Good for Equality and Good for Business." *Legal Intelligencer,* July 3. Retrieved from www.blankrome.com/publications/understanding -unconscious-bias-good-equality-and-good-business.

Levere, Jane L. 2018. "Role Models Tell Girls That STEM's for Them in New Campaign." *New York Times,* September 9, 2018. www.nytimes .com/2018/09/09/business/media/ad-council-stem-girls.html.

Lewis, Jane, and Mary Campbell 2008. "What's in a Name? 'Work and Family' or 'Work and Life' Balance Policies in the UK since 1997 and the Implications for the Pursuit of Gender Equality." *Social Policy and Administration* 42:524–41.

Lichtenstein, Nelson. 2003. *State of the Union.* Princeton, NJ: Princeton University Press.

Liou, Yu-Ming, and Paul Musgrave. 2016. "Oil, Autocratic Survival, and the Gendered Resource Curse: When Inefficient Policy Is Politically Expedient." *International Studies Quarterly* 60(3):440–56.

Looze, Jessica. 2017. "Why Do(n't) They Leave? Motherhood and Women's Job Mobility." *Social Science Research* 65:47–59.

Lorde, Audre. 1984. *Sister Outsider: Essays and Speeches.* Trumansburg, NY: Crossing.

Maddow, Rachel. 2019. *Blowout: Corrupted Democracy, Rogue State Russia, and the Richest, Most Destructive Industry on Earth.* New York: Crown.

McColgan, Paul. 2019. "The Damage of Relentless Layoffs in the Oil and Gas Industry." LinkedIn, May 19. www.linkedin.com/pulse /damage-relentless-layoffs-oil-gas-industry-paul-mccolgan-ph-d-.

McLaughlin, Heather, Christopher Uggen, and Amy Blackstone. 2017. "The Economic and Career Effects of Sexual Harassment on Working Women." *Gender and Society* 31(3):333–58.

Melosi, Martin. 1995. "Equity, Eco-Racism, and Environmental History." *Environmental History Review* 19(3):1–16.

Metcalf, Heather. 2010. "Stuck in the Pipeline: A Critical Review of STEM Workforce Literature." *Interactions: UCLA Journal of Education and Information Studies.* Online resource. https://escholarship.org/uc/item/6zf09176.

Mikati, Ihab, Adam F. Benson, Thomas J. Luben, Jason D. Sacks, and Jennifer Richmond-Bryant. 2018. "Disparities in Distribution of Particulate Matter Emission Sources by Race and Poverty Status." *American Journal of Public Health* 108(4):480–85. https://doi.org/10.2105/AJPH.2017.304297.

Mills, C. Wright. 1959. *The Sociological Imagination.* New York: Oxford University Press.

Moore, Wendy Leo, and Joyce M. Bell. 2011. "Maneuvers of Whiteness: 'Diversity' as a Mechanism of Retrenchment in the Affirmative Action Discourse." *Critical Sociology* 37:597–613.

Mundy, Liza. 2017. "Why Is Silicon Valley so Awful to Women?" *Atlantic,* April.

Myatt, Mike. 2012. "Ten Reasons Your Top Talent Will Leave You." *Forbes,* December 13. www.forbes.com/sites/mikemyatt/2012/12/13/10-reasons-your-top-talent-will-leave-you/.

NAACP (National Association for the Advancement of Colored People). 2019. Environmental and Climate Justice Program. https://naacp.org/environmental-climate-justice-about/.

National Academies. 2010. *Rising above the Gathering Storm, Revisited.* Washington, DC: National Academies Press.

NCSES (National Center for Science and Engineering Statistics). 2019. "S and E Graduate Students, by Field, Sex, Citizenship, Ethnicity, and Race: 2016." NSF 19–304. https://ncses.nsf.gov/pubs/nsf19304/data.

Neely, Megan Tobias. 2020. "The Portfolio Ideal Worker: Insecurity and Inequality in the New Economy." *Qualitative Sociology* 43:271–96.

Nelson, Robin G., Julienne N. Rutherford, Katie Hinde, and Kathryn B. H. Clancy. 2017. "Signaling Safety: Characterizing Fieldwork Experiences and Their Implications for Career Trajectories." *Ameri-*

can *Anthropologist* 119(4):710–22. https://anthrosource.onlinelibrary
.wiley.com/doi/pdf/10.1111/aman.12929.

Neuhauser, Alan. 2018. "Oil Boom a Bust for Blacks." *US News and World Report,* August 24.

Norris, Dawn R. 2016. *Job Loss, Identity, and Mental Health.* New Brunswick, NJ: Rutgers University Press.

NSF (National Science Foundation). 2017. "ADVANCE: Increasing the Participation and Advancement of Women in Academic Science, Technology, Engineering and Mathematic Careers." NSF publication 17–70.

NSF (National Science Foundation). 2018. "ADVANCE: Organizational Change for Gender Equity in STEM Academic Professions." www.nsf.gov/funding/pgm_summ.jsp?pims_id=5383.

Osnowitz, Debra. 2010. *Freelancing Expertise: Contract Professionals in the New Economy.* Ithaca, NY: ILR.

Perks, Rachel, and Katrin Schulz. 2020. "Gender in Oil, Gas and Mining: An Overview of the Global State-of-play." *Extractive Industries and Society* 7:380–88.

Pettit, Becky, and Jennifer Hook. 2009. *Gendered Tradeoffs: Family, Social Policy, and Economic Inequality in 21 Countries.* New York: Russell Sage Foundation.

Pollack, Eileen. 2015. *The Only Woman in the Room: Why Science Is Still a Boy's Club.* Boston: Beacon.

Powell, Walter W. 2001. "The Capitalist Firm in the 21st Century: Emerging Patterns in Western Enterprise." In *The Twenty-first Century Firm,* edited by P. DiMaggio. Princeton, NJ: Princeton University Press.

Priest, Tyler, and Michael Botson. 2012. "Bucking the Odds: Organized Labor in Gulf Coast Oil Refining." *Journal of American History,* June.

Pugh, Allison J. 2013. "What Good Are Interviews for Thinking about Culture? Demystifying Interpretive Analysis." *American Journal of Cultural Sociology* 1:42–68.

Pugh, Allison J. 2015. *The Tumbleweed Society: Working and Caring in an Age of Insecurity.* Oxford: Oxford University Press.

Rao, Aliya Hamid. 2020. *Crunch Time: How Married Couples Confront Unemployment.* Oakland: University of California Press.

Rao, Aliya Hamid, and Megan Tobias Neely. 2019. "What's Love Got to Do with It? Passion and Inequality in White-collar Work." *Sociology Compass*. https://doi.org/10.1111/soc4.12744.

Ray, Victor. 2019. "A Theory of Racialized Organizations." *American Sociological Review* 84(1):26–53.

Rick, Katharina, Iván Martén, and Ulrike Von Lonski. 2017. "Untapped Reserves: Promoting Gender Balance in Oil and Gas." World Petroleum Council and Boston Consulting Group. www.bcg .com/publications/2017/energy-environment-people-organization -untapped-reserves.

Riegle-Crumb, Catherine, and Melissa Humphries. 2012. "Exploring Bias in Math Teachers' Perceptions of Students' Ability by Gender and Race/Ethnicity." *Gender and Society* 26(2):290–322. doi: 10.1177 /0891243211434614.

Riegle-Crumb, Catherine, and Chelsea Moore. 2014. "The Gender Gap in High School Physics: Considering the Context of Local Communities." *Social Science Quarterly* 95(1):253–68.

Roscigno, Vincent, Sherry Mong, Reginald Byron, and Griff Tester. 2007. "Age Discrimination, Social Closure and Employment." *Social Forces* 86(1):313–34.

Ross, Michael L. 2012. *The Oil Curse: How Petroleum Wealth Shapes the Development of Nations*. Princeton, NJ: Princeton University Press.

Rousseau, Denise, Violet T. Ho, and Jerald Greenberg. 2006. "I-Deals: Idiosyncratic Terms in Employment Relationships." *Academy of Management Review* 31(4): 977–94.

Sandberg, Sheryl. 2013. *Lean In: Women, Work, and the Will to Lead*. New York: Knopf.

Scott, David, and KangJae Jerry Lee. 2018. "People of Color and Their Constraints to National Park Visitation." *George Wright Forum* 35(1): 73–82.

Scott, Jen, Rose Dakin, Katherine Heller, Adriana Eftimie. 2013. "Extracting Lessons on Gender in the Oil and Gas Sector: A Survey and Analysis of the Gendered Impacts of Onshore Oil and Gas Production in Three Developing Countries." World Bank Oil, Gas, and Mining Unit Working Paper.

Sharone, Ofer. 2013. *Flawed System/Flawed Self.* Chicago: University of Chicago Press.

Shettle, Carolyn F. 1997. *Who Is Unemployed? Factors Affecting Unemployment among Individuals with Doctoral Degrees in Science and Engineering.* Washington, DC: National Science Foundation. https://files.eric .ed.gov/fulltext/ED417084.pdf.

Smith, Dorothy. 1979. "A Sociology for Women." In *The Prism of Sex: Essays in the Sociology of Knowledge.* Madison: University of Wisconsin Press.

Spence, Mark David. 1999. *Dispossessing the Wilderness: Indian Removal and the Making of National Parks.* New York: Oxford University Press.

Stacey, Judith. 1998. *Brave New Families.* Berkeley: University of California Press.

Stokes, Philip J., Roger Levine, and Karl W. Flessa. 2015. "Choosing the Geoscience Major: Important Factors, Race/Ethnicity, and Gender." *Journal of Geoscience Education* 63(3):250–63. doi: 10.5408/14–038.1.

Stone, Pamela. 2007. *Opting Out? Why Women Really Quit Careers and Head Home.* Berkeley: University of California Press.

Subramaniam, Banu. 2009. "Moored Metaphorphoses: A Retrospective Essay on Feminist Science Studies." *Signs* 34(4):951–80.

Taylor, Dorceta E. 2014. *Toxic Communities: Environmental Racism, Industrial Pollution, and Residential Mobility.* New York: NYU Press.

Taylor, Dorceta E. 2016. *The Rise of the American Conservation Movement: Power, Privilege, and Environmental Protection.* Durham, NC: Duke University Press.

Teitelbaum, Michael S. 2014. *Falling Behind? Boom, Bust, and the Global Race for Scientific Talent.* Princeton, NJ: Princeton University Press.

Thébaud, Sarah, and Maria Charles. 2018. "Segregation, Stereotypes, and STEM." *Social Sciences* 7 (111). doi:10.3390/socsci7070111 www .mdpi.com/journal/socsci.

Thistle, Susan. 2006. *From Marriage to the Market.* Berkeley: University of California Press.

Torres, Nicole. 2017. "The H-1B Visa Debate, Explained." *Harvard Business Review,* May 4. https://hbr.org/2017/05/the-h-1b-visa-debate-explained.

UNECE (United Nations Economic Commission for Europe). 2019. *Promoting Gender Diversity and Inclusion in the Oil, Gas and Mining*

Extractive Industries: A Woman's Human Rights Report. Minneapolis, MN: Advocates for Human Rights. www.unece.org/fileadmin /DAM/energy/images/CMM/CMM_CE/AHR_gender_diversity _report_FINAL.pdf.

US Bureau of Labor Statistics. 2012. *Employed Persons by Detailed Industry, Sex, Race, and Hispanic or Latino Ethnicity.* Washington, DC: Bureau of Labor Statistics. www.bls.gov/cps/cpsaat18.pdf.

Vallas, Steven. 2011. *Work: A Critique.* Boston: Polity.

Vallas, Steven, and Elaine Cummins. 2015. "Personal Branding and Identity Norms in the Popular Business Press: Enterprise Culture in an Age of Precarity." *Organization Studies* 36:293–319.

Wacquant, Loïc. 1990. "Sociology as Socioanalysis." *Sociological Forum* 5(4):677–89.

Wald, Ellen. 2018. *Saudi, Inc.: The Arabian Kingdom's Pursuit of Profit and Power.* New York: Pegasus.

Webber, Gretchen, and Christine L. Williams. 2008. "Mothers in 'Good' and 'Bad' Part-time Jobs: Different Problems, Same Results." *Gender and Society* 22:752–77.

Weeden, Kim A. 2002. "Why Do Some Occupations Pay More than Others? Social Closure and Earnings Inequality in the United States." *American Journal of Sociology* 108(1):55–101.

Williams, Christine L. 2017. "The Gender of Layoffs in the Oil and Gas Industry." *Research in the Sociology of Work* 31:215–42.

Williams, Christine L. 2019. "The Deserving Professional: Job Insecurity and Gender Inequality in the Oil and Gas Industry." *Labour and Industry* 29(2):199–212. www.tandfonline.com/doi/full/10.1080/10 301763.2019.1600856.

Williams, Christine L. 2021. "Jessie Bernard, Feminism, and 'Her' Marriage to Luther Bernard." *Sociological Forum* 36(2):524–27.

Williams, Christine L., Kristine Kilanski, and Chandra Muller. 2014. "Corporate Diversity and Gender Inequality in the Oil and Gas Industry." *Work and Occupations* 41:440–76.

Williams, Christine L., Chandra Muller, and Kristine Kilanski. 2012. "Gendered Organizations in the New Economy." *Gender and Society* 26: 549–73.

Williams, Joan. 2000. *Unbending Gender: Why Family and Work Conflict and What to Do about It.* New York: Oxford University Press.

Williams, Joan, and Rachel Dempsey. 2014. *What Works for Women at Work.* New York: NYU Press.

Wilson, Carolyn. 2016. *Status of the Geoscience Workforce 2016.* Alexandria, VA: American Geosciences Institute.

Wingfield, Adia Harvey. 2009. "Racializing the Glass Escalator: Reconsidering Men's Experiences with Women's Work." *Gender and Society* 23(1):5–26.

Wingfield, Adia Harvey. 2010. "Are Some Emotions Marked 'Whites Only'? Racialized Feeling Rules in Professional Workplaces." *Social Problems* 57:251–68.

Wingfield, Adia Harvey. 2013. *No More Invisible Man: Race and Gender in Men's Work.* Philadelphia: Temple University Press.

Wingfield, Adia Harvey. 2019. *Flatlining: Race, Work, and Health Care in the New Economy.* Oakland: University of California Press.

Wingfield, Adia Harvey, and Renée Skeete Allston. 2014. "Maintaining Hierarchies in Predominantly White Organizations: A Theory of Racial Tasks." *American Behavioral Scientist* 58(2):274–87.

Yergin, Daniel. 1991. *The Prize: The Epic Quest for Oil, Money, and Power.* New York: Simon and Schuster.

Yergin, Daniel. 2011. *The Quest: Energy, Security, and the Remaking of the Modern World.* New York: Penguin Random House.

ILLUSTRATION CREDITS

Figure 1.   Delegates at the 2018 OPEC. Used with permission from
OPEC.

Figure 4.   CEOs of the US largest oil companies testifying before the
US Congress. Photo: Doug Mills/*The New York Times*.
Used with permission from Xinhua/eyevine/Redux.

Figure 5.   Crude Oil Prices. Source: www.macrotrends.net/1369
/crude-oil-price-history-chart.

Figure 9.   Images from the "Bearded Lady Project." Photographed by
Kelsey Vance. Copyright The Bearded Lady Project 2019.
Used with permission.

Figure 10. The iconic geologist at the Smithsonian Natural History
Museum. Used with permission from the Grove National
Historic Landmark Archives, Glenview Park District.

INDEX

Italicized page numbers indicate figure locations.